Imaginative Ideas for the Teacher of Mathematics, Grades K–12
RANUCCI'S RESERVOIR

A Collection of Articles by Ernest R. Ranucci

Edited by
Margaret A. Farrell
University at Albany
State University of New York

NATIONAL COUNCIL OF TEACHERS OF MATHEMATICS

Copyright © 1988 by
THE NATIONAL COUNCIL OF TEACHERS OF MATHEMATICS, INC.
1906 Association Drive, Reston, Virginia 22091
All rights reserved

Library of Congress Cataloging in Publication Data:

Ranucci, Ernest R.
    Imaginative ideas for the teacher of mathematics, grades K–12.

    Bibliography: p.
    1. Mathematics—Study and teaching.   I. Farrell, Margaret A. II. Title. III. Title: Ranucci's reservoir.
QA12.R36 1988   372.7   88-12576
ISBN 0-87353-257-0

The publications of the National Council of Teachers of Mathematics present a variety of viewpoints. The views expressed or implied in this publication, unless otherwise noted, should not be interpreted as official positions of the Council.

Printed in the United States of America

Table of Contents

Introduction: Biography of Ernest R. Ranucci 1

Part 1: Patterns

Alphabet Soup 4
    (September 1968 *Jack and Jill*)
Patterns in Figures 5
    (September 1963 *Instructor*)
4 Areas in the New Math 7
    (February 1964 *Instructor*)
Mathematics on the Ceiling 14
    (*Updating Mathematics: Junior High*, 1964)
The Calculus of Finite Differences 18
    (*Professional Growth for Teachers*, 1965)
Discovery in Mathematics 21
    (January 1965 *Arithmetic Teacher*)
Function Follows Form 26
    (April 1966 *Arithmetic Teacher*)

Part 2: Mathematics in the World

Music in the Marshall Islands 32
    (April 1973 *School Science and Mathematics*)
Of Shoes—and Ships—and Sealing Wax—of Barber Poles and Things 36
    (April 1975 *Mathematics Teacher*)

Part 3: And Then There Was Space!

The Weequahic Configuration 42
    (February 1960 *Mathematics Teacher*)
Drawing for Math Teachers Who Can't Draw 45
    (*Updating Mathematics: High School*, 1964)
Schlegel Diagrams 49
    (April 1971 *Journal of Recreational Mathematics*)
Spatial Aspects of the Venn Diagram 57
    (1968 *New York State Mathematics Teachers Journal*)
Topology—through the Alphabet 60
    (December 1972 *Mathematics Teacher*)

Part 4: Inventiveness in Geometry

On Skewed Regular Polygons 64
    (March 1970 *Mathematics Teacher*)
The Congruency of Quadrilaterals 67
    (September 1973 *Mathematics Teaching*)
A Tiny Treasury of Tessellations 70
    (February 1968 *Mathematics Teacher*)
Master of Tessellations: M. C. Escher, 1898–1972 74
    (April 1974 *Mathematics Teacher*)

Part 5: Games to Learn By

What's in a Name? 82
    (January 1970 *Grade Teacher*)
Tantalizing Ternary 84
    (December 1968 *Arithmetic Teacher*)
Dots and Squares 88
    (January 1969 *Journal of Recreational Mathematics*)

Selected Bibliography of Works by Ernest R. Ranucci 91

Introduction

Biography of Ernest R. Ranucci

Ernest R. Ranucci was born in Newark, New Jersey, on 26 July 1912. He died at his summer home in Brookfield, Vermont, in August of 1976. His bachelor's and master's degrees were in mathematics, with a second field in science, from Montclair State Teachers College in 1933 and 1938, respectively. His Ph.D. in mathematics education was granted by Columbia University in 1952.

For almost thirty years, he taught secondary school mathematics and supervised secondary and college mathematics departments in schools in New Jersey. The last of these was the mathematics department at Newark State College in Newark, New Jersey, where he served as department chair from 1957 to 1963. In 1953, he was awarded the first of several Fulbright fellowships—in this instance to teach and study during the academic year in Dundee, Scotland. His interest in mathematics education from an international perspective grew steadily. In 1959 and 1960 he was an exchange teacher in San Salvador, El Salvador. In 1963 and 1964 he acted as director of a graded school in São Paulo, Brazil. Subsequently, he acted as coordinator of mathematics in Brazil, Uruguay, Argentina, Chile, Peru, Ecuador, Bolivia, and Paraguay. He taught classes in Micronesia to Peace Corps teachers, traveled to the United Nations school in New York City to teach first graders once a week, and participated in a teacher training program in Costa Rica.

In 1965, after his return from a year in Brazil, Ranucci joined the faculty of the State University of New York at Albany as a professor of mathematics education in the Department of Instruction. At Albany, he designed and taught a mathematics course in advanced Euclidean geometry, brought leading scholars to the campus for full-day lectures to area mathematics teachers, and soon had an enthusiastic following of undergraduate and graduate students in mathematics education who wanted to "take a course from Ranucci."

He was a popular speaker at professional meetings and loved to be given a title like "Geometry and the Imagination" so that he could mold his talk in as creative a fashion as possible. Both he and his wife, Barbara, were regular participants at the summer workshops for mathematics teachers held by the Association of Mathematics Teachers in New England. In the evenings, they had an opportunity to lead participants in another Ranucci passion, music.

Ernest Ranucci's writing was as much a part of his life as breathing. The maxim "publish or perish" was meaningless to him. He wrote in order to reach a wider audience, to crystallize half-formed ideas, and because he loved to write. His record of over 110 articles, 5 books, and numerous clips in periodicals like *Jack and Jill* or *Reader's Digest* is the formal record of his effort. His students know that, in addition, he created volumes of loose-leaf handouts of teaching ideas, problem situations, and drawings. Writing was so important to Ranucci as a way of teaching and learning that he encouraged his graduate students to submit manuscripts to journals as part of their course work. He willingly provided the first level of editorial comment and helped these novice authors choose appropriate journals.

Ernest Ranucci loved geometry; he was particularly interested in spatial perception, his area of doctoral research, and he loved teaching at all levels. He was an active contributor to the profession of mathematics education. He had been one of the members on the national panel of the famous 1959 Commission on Mathematics Report. Throughout his career, Ranucci contributed his expertise as a speaker at the regional and annual meetings of the National Council of Teachers of Mathematics and at state and regional meetings—especially in New York and New

England. He was an invited speaker at the 1972 Budapest meeting of the International Colloquium on Theoretical Problems in the Teaching of Mathematics in the Primary Schools. Ernest Ranucci never missed an opportunity to inspire others with the beauty of mathematics. Perhaps the most striking example of this trait was his spontaneous visits to elementary schools as he traveled across the United States. He would introduce himself to the principal and ask if he could visit a classroom. Soon this eternally young teacher would be teaching an enthusiastic group of second graders.

As readers of this book will discover for themselves, Ernest Ranucci's ideas are as vital and applicable today as they were when he wrote them.

Part 1

Patterns

The six articles selected for the first section of this book have one common characteristic. All emphasize patterns and ways of encouraging students to find patterns, both geometric and numerical. In "Alphabet Soup," the young readers of *Jack and Jill* are encouraged to observe shape and orientation commonalities in order to find "the next letter." "Patterns in Figures," from the *Instructor,* contains sets of exercises that move from geometric patterns to numerical patterns and back again. The use of "What comes next?" activities to help students develop their inductive reasoning abilities was a common thread through Ranucci's writings and teachings. In "4 Areas in the New Math," from the *Instructor,* he used pattern generation as a problem-solving technique and geometric shape as a way of giving meaning to arithmetic concepts. Don't be put off by the "New Math" part of the the title; the ideas are just as applicable in today's problem-solving curriculum. Even more important, all three articles contain teaching ideas that transcend the age group usually associated with *Jack and Jill* and taught by most readers of the *Instructor.*

The next two readings in Part 1 are from source books for teachers. In "Mathematics on the Ceiling," Ranucci used ceiling tiles (a favorite hands-on manipulative of his) to lead the reader through applications of the binomial theorem and then into a little lattice theory. Although there have been many recent articles, typically using the geoboard, that have developed area on a lattice, Ranucci did this in a unique way—he shared his thinking processes with the reader. The next reading, "The Calculus of Finite Differences," is misnamed "Enrichment. . . ." In this article, Ranucci described a powerful analytical tool for use in algebra classes and an elegant precursor to the use of the derivative in calculus. In his introduction, Ranucci differentiated problem types solved by experimental means from those accessible by generalization—the latter being the subject of the article.

In "Discovery in Mathematics," from the *Arithmetic Teacher,* the bowling pin illustration seen in the preceding reading reappears. This time, observation, different ways to list information systematically, and the use of arithmetic no more sophisticated than counting are emphasized. Meanings for labels such as *permutations* and *isomorphic* are provided for the teacher. (Science purists may object to the way in which the word *experiment* is used here.) As Ranucci's comments made clear, the important part of the article is the activities that were used successfully with children in the intermediate grades. The final article in this section, "Function Follows Form," from the *Arithmetic Teacher,* continues the theme of pattern generation with emphasis on the geometry of form and its connections with the number world. Here we find the kind of shape relationships recognized by the Pythagoreans, for example, triangular numbers and square numbers. Even the classic handshake problem is modeled in a geometric way, in terms of form. Ranucci is making an important point here. It is not enough to help children find patterns; it is especially important to recreate some of those patterns in visual or geometric format. Ranucci mentioned Piaget in his introduction to the article. Now, over twenty years later, more recent research on developmental approaches in instruction would as strongly support Ranucci's emphasis on form.

Alphabet Soup

By ERNEST R. RANUCCI

The letters in each separate line below are arranged in a logical pattern. Think about the pattern, and try to figure out what the next letter should be. You will see that some of the letters are lying down or leaning over; their position is important. In one line, the shape of the letters is important. This is an exercise in logic that is tricky but fun. We've done the first one. Can you figure out the rest?

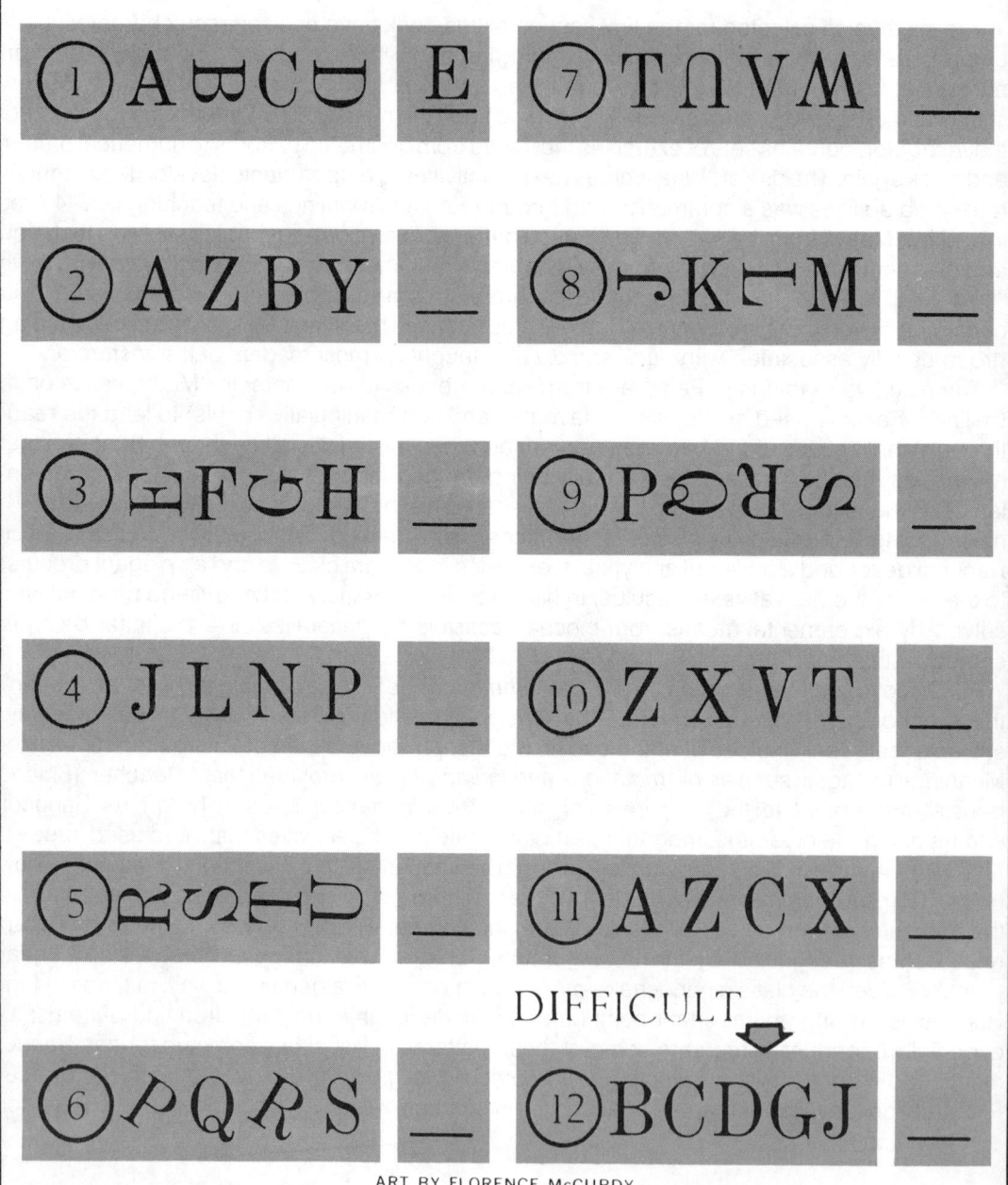

ART BY FLORENCE McCURDY

Answers on page 83

Reprinted by permission of INSTRUCTOR, September 1963. Copyright © 1963 by the Instructor Publications, Inc.

What Comes Next?
Patterns in Figures

ERNEST R. RANUCCI — Coordinator of Mathematics, The American Elementary and High School, Sao Paulo, Brazil

CAN you tell what comes next in each of the following patterns? Supply the next number, symbol, or drawing, as the case may be. (Some of the patterns could go on and on, but give just the next one except where there are dashes to fill in.) The answers are on page 83, but don't peek at them too soon! P.S. You might like to make up some patterns of your own.

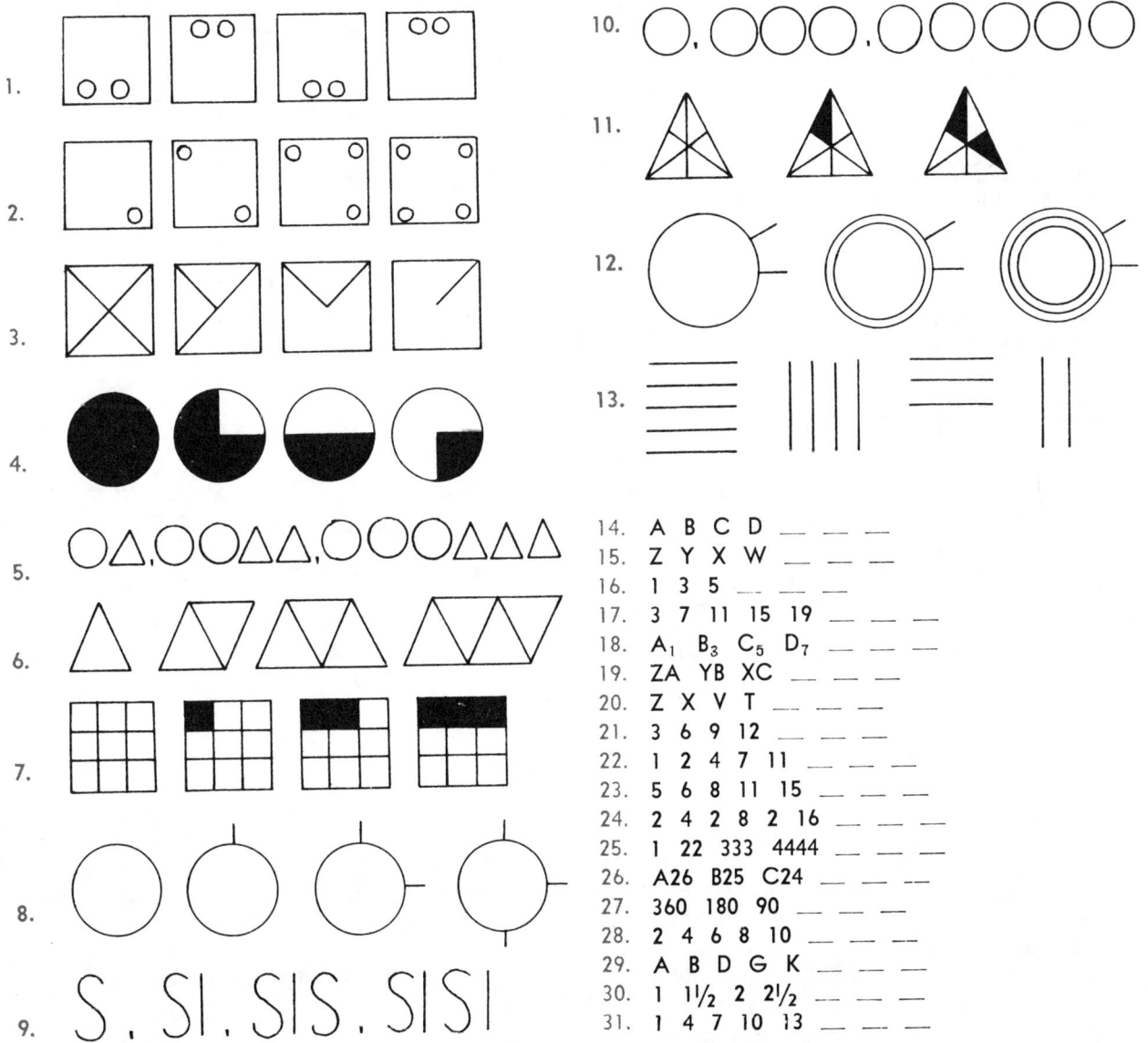

14. A B C D __ __ __
15. Z Y X W __ __ __
16. 1 3 5 __ __ __
17. 3 7 11 15 19 __ __
18. A_1 B_3 C_5 D_7 __ __
19. ZA YB XC __ __
20. Z X V T __ __ __
21. 3 6 9 12 __ __
22. 1 2 4 7 11 __ __
23. 5 6 8 11 15 __ __
24. 2 4 2 8 2 16 __ __
25. 1 22 333 4444 __ __
26. A26 B25 C24 __ __
27. 360 180 90 __ __ __
28. 2 4 6 8 10 __ __
29. A B D G K __ __
30. 1 1½ 2 2½ __ __
31. 1 4 7 10 13 __ __

Patterns

5

32. ½ ¼ 1/6 1/8 ___ ___
33. A N B O C P ___ ___
34. 5 15 45 135 ___ ___
35. 3 7 15 31 ___
36. A Z B X C V ___ ___
37. 1 3 2 4 3 5 ___ ___
38. A C E G ___
39. B A D C F E ___ ___
40. 1 2 1 1/16 1 15/16 ___
41. 1 4 27 256 ___
42. STORE, TORE, ORE, RE ___
43. ½ 3/4 6/7 10/11 ___ ___
44. boy, three, crowd, five, jam, ___
45. eat, fat, gat, ___
46. AMZ BNY COX ___ ___
47. 6 oz.; 13 oz.; 1 lb. 4 oz.; ___
48. HOLLOW, OLLOW, OLLO, LLO, ___
49. A_1 Z_{26} B_3 Y_{24} C_5 X_{22} ___ ___
50. 31 28 31 30 31 30 31 ___ ___
51. 1 1 2 6 24 120 ___ ___
52. !? !!?? ___ ___
53. 3 8 13 18 23 ___ ___
54. 5 6 8 11 15 ___
55. A Z B Y C X ___ ___
56. A Z B X C V ___ ___
57. Z W T Q ___ ___
58. 6 in.; 1 ft. 1 in.; 1 ft. 8 in.; ___
59. ABCD GHIJ MNOP ___
60. 1 2 3 5 8 13 21 ___
61. 2 4 2 8 2 16 ___ ___
62. 1 4 3 6 5 8 7 ___ ___
63. A L B M C N D ___ ___
64. 3600 1800 600 150 ___ ___

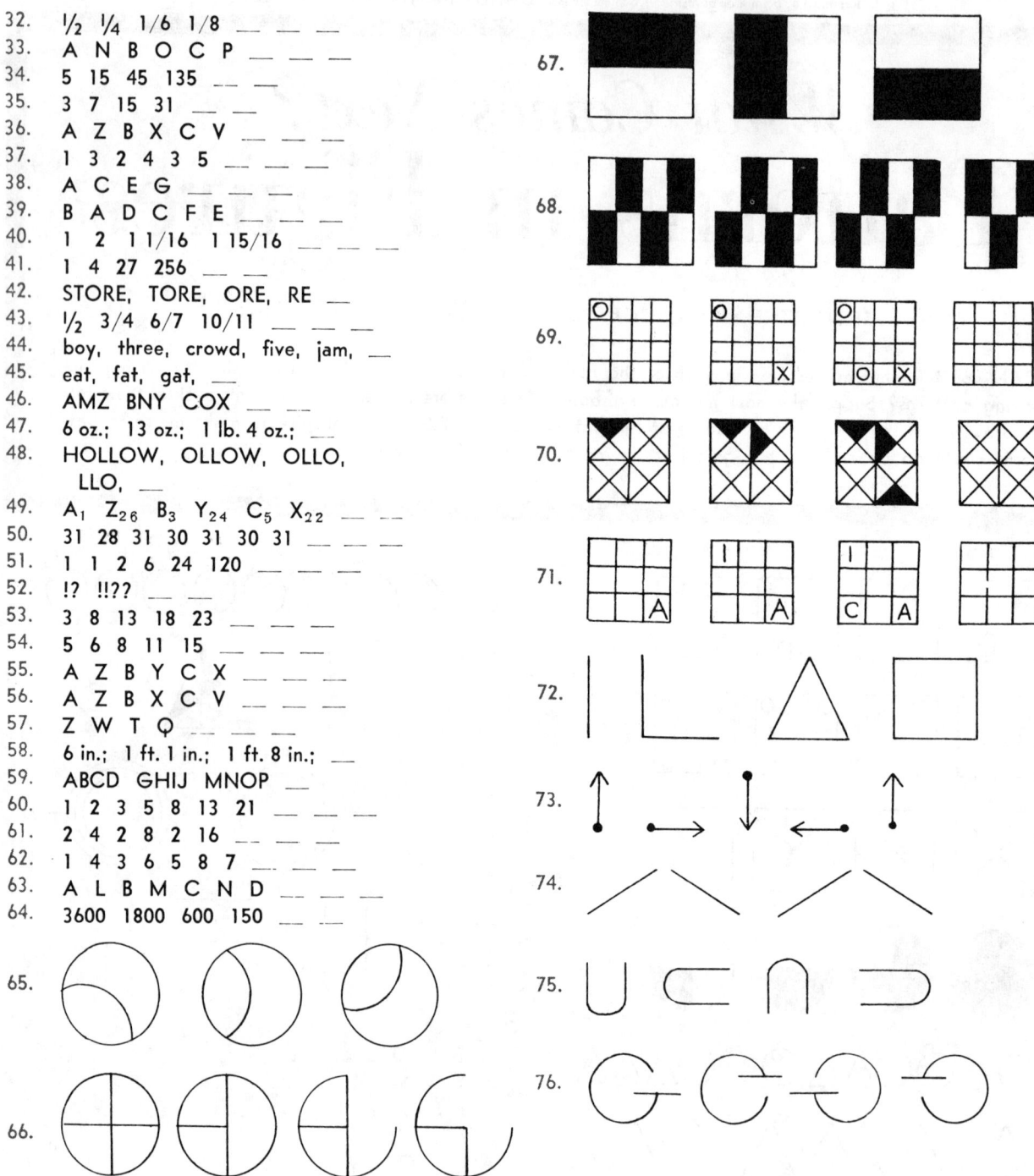

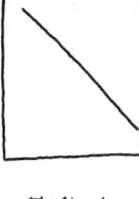

Draw a short line only partly across the sheet of paper. If you color one side of the short line red, you'll have to color the entire paper red. You won't know when to stop coloring red.

Draw a triangle in the center of a sheet of paper, leaving part of one side missing. If you color the outside of the drawing blue, you'll have to color the inside blue. You won't know when the outside becomes the inside.

In the first two diagrams above, the drawing is said to *separate* the sheet of paper. In the last two, the drawing does *not* separate the sheet of paper. In the following sketches, you decide whether the surface is separated or not. Arrows placed at the end of a short line segment mean that this line continues on and on. If no arrows are placed, the segment stops at its ends; sometimes we indicate this stopping by placing a large dot at each end of such a segment.

Certain natural laws govern the separation or non-separation of all drawings. The next series of experiments will develop such laws. But we shall need to know more about *points*, *regions*, and *segments* before we are able to do these experiments.

4 Areas in the New Math

ERNEST R. RANUCCI
Coordinator of Mathematics
The American Elementary and High School, Sao Paulo, Brazil

Points, Regions, and Segments

Draw a line completely across a sheet of paper. You could color one side red. You could color the other side green. You could use any colors you cared to. The fact remains that the line would separate two different colors.

Draw a circle in the center of a sheet of paper. You could color the inside of the circle blue. You could color the outside of the circle red. You could use any colors you cared to. The fact remains that the line would separate two different colors.

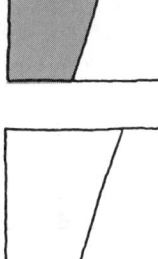

Patterns

7

Readers may now have another question. Do relations similar to those just discussed govern the behavior of three-dimensional objects? Are there additional geometric terms which will have to be learned?

Both *point* and *segment* mean, in three dimensions, about what they mean in two dimensions. However, the word *point* is usually replaced by the word *vertex* (plural *vertices*). The word *segment* is commonly replaced by *edge*. An additional term, *face*, is usually used in describing three-dimensional objects. In the cube, for example, there are eight corners (*points*) or *vertices*. There are twelve *edges* and six *faces*. An Egyptian pyramid has five *vertices*, five *faces*, and eight *edges*. Some three-dimensional objects are sketched below. The table summarizes the appropriate entries. Notice that the entry in the E column seems to be 2 less than the sum of the entries in columns F and V. This may be indicated by the formula $F + V = E + 2$. Known as Euler's Formula for three-dimensional objects, it is just about identical in spirit to the formula explaining two-dimensional relations.

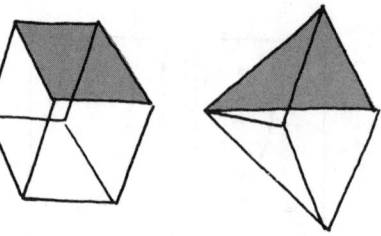

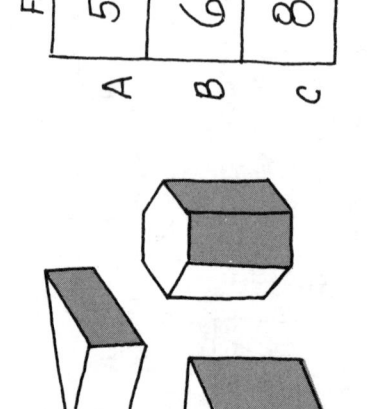

	F	V	E
A	5	6	9
B	6	8	12
C	8	12	18

Euler's Formula lends itself to the following activity. Collect a series of wooden blocks in a variety of forms and shapes. The only requirement is that the faces of the blocks be flat. Count the number of faces, vertices, and edges of each block. By working in

---

A *point* fixes a position. It has neither length nor width nor depth. Think of it as the end of the sharpest needle ever made.

A *segment* is a collection of points with a definite beginning and a definite end. In some cases the segment will have the same beginning and end; think of a snake biting its own tail.

A *region* is formed whenever separation occurs. There are three regions in the sketch at the right. If separation does not occur, the whole surface or plane is considered one region.

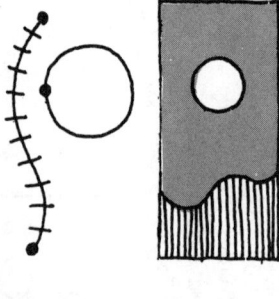

	P	R	S
·	1	1	0
•—•	2	1	1
	2	2	2
	3	2	3
	4	2	4
	4	3	5
	4	4	6
	4	5	7
	4	6	8

The three columns labeled P, R, and S in the chart at the right represent, respectively, Points, Regions, and Segments. The number of each of these geometric forms is entered on the chart, as a fairly complicated drawing is being gradually built up.

As the drawing progresses, observe the bookkeeping entries. Do you notice a relationship between the entries in each row? The sum of the numbers in P and R appears to be two larger than the entry in S. This may be indicated by the formula $P + R = S + 2$. It is not actually known who first discovered this relation but credit is usually given to the Swiss mathematician Euler. Every drawing similar in nature to these conforms to this formula. Analyze several drawings to check this for yourself.

Imaginative Ideas: Ranucci's Reservoir

pairs, you and your neighbor may confirm each other's bookkeeping. Can you show for yourself the Euler relation?

Ask yourself these questions. What is the fewest number of faces which may meet to form a vertex? What is the fewest number of colors which may be used to distinguish the six faces of a cube? What is the fewest number of colors which may be used to distinguish the five faces of an Egyptian pyramid? Is is possible to cut through a wooden cube in such a manner as to get a cross section with six edges? If a cube is three inches long, how many times must it be sliced to form one-inch cubes? How many such small cubes will be formed? How does the flat pattern of a cube look?

Experiments with Probability

One of the most refreshing developments in mathematics in the last twenty-five years has been the introduction of simple aspects of serious mathematical topics into the curricula of the elementary school. Among the more stimulating of these topics has been the field of probability.

Coin Spinning

Your dialogue might be something like this: "I have a coin here. It has a head and a tail. When I spin the coin, it doesn't know whether to fall head or tail. It just falls and something called luck or chance or probability makes it fall one way or the other. If I spun this coin one million times, it would certainly surprise me to find that it had fallen heads up one million times. It would, of course, surprise me to find it had fallen tails up one million times. Somewhere between these two extremes I would not be specially surprised one way or the other. Suppose that we take this coin. I shall spin it one hundred times. Write on your paper the number of times you think it will fall heads and the number of times that it will fall tails. Now we'll actually spin the coin and find out just how it behaves."

From experiments such as these, children arrive at the intuitive notion that, since the coin can fall only one way or the other, there is a fifty-fifty chance of falling heads and a fifty-fifty chance of falling tails. The idea that the probability of falling heads ($\tfrac{1}{2}$) and the probability of falling tails ($\tfrac{1}{2}$) is the same, is now introduced. Children are interested in the "runs" which occur; they will enthusiastically cheer when either heads or tails have been having a hard time.

Children will be interested in the possibility of adding the results of their experiment to those of other classes. This is valuable since the results of tossing a coin only one hundred times may be somewhat inconclusive. One fourth-grade group was quite interested to learn that in three other classes tails had been falling much too frequently. They were *very* satisfied when heads fell 70 times and tails 30 times in their experiment with one hundred tosses.

Another significant experiment in probability is for everyone to perform the experiment at home. The number of heads and tails resulting from the same experiment performed by 25 or 30 students is likely to be more significant than a single experiment. It is interesting also to see which of the students achieve results closest to those expected and farthest from those expected.

Out of experiments of this type should emerge certain mathematical "truths" or guidelines.

The probability that a coin will fall heads and the probability that a coin will fall tails are equally likely with an honest coin. We may state this economically in the form: Probability of Heads (P$_H = \tfrac{1}{2}$); Probability of Tails (P$_T = \tfrac{1}{2}$).

The probability that a coin will fall heads and the probability that a coin will fall tails must have a sum of one. Similarly, if the probability that a certain event will occur is $\tfrac{1}{3}$, the probability that it will *not* occur is $\tfrac{2}{3}$. This means that were the event to occur 3,000 times, about 1,000 times it would be successful and 2,000 times it would fail.

When a thing MUST happen, the probability is said to be 1. When a thing CANNOT happen, the probability is said to be 0.

Number Line Activities

In the *24th Yearbook* of the National Council of Teachers of Mathematics, David A. Page has a series of exercises performed on a type of number line. A line segment perhaps ten inches long is drawn and divided into ten equal parts, as below—the division points are marked in tenths from 0 to 1.

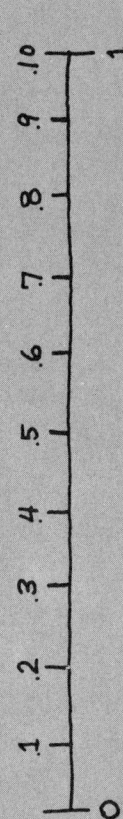

A series of questions is asked, and an estimate of the likelihood of occurrence of each event is marked on the segment. In question A, for example, an A is placed at the .5 point on the probability line. An honest coin is tossed once. What is the probability that it will fall heads? (½)

D. What is the probability that it will fall tails? (½)
   A box contains four marbles, three whites and one black. If you reach in once, what is the probability that you will pick the black one? (¼)
E. What is the probability that you will pick a white one? (¾)
F. What is the probability that at exactly 2 P.M. today, the door to the classroom will open and a man from Mars with seventeen legs and three eyes will enter? (Close to 0, we hope!)
   What is the probability that you will be promoted at the end of the school year? (Probability may vary widely.)
G. There are ten books on my desk. I am thinking of one of them, not necessarily the mathematics book; what is the probability that you will guess the one that I am thinking of? (¹⁄₁₀)
H. What is the probability that at exactly 10 A.M. today, lightning will strike the center desk in this room? (Close to 0)
I. What is the probability that you will be alive 1,000 years from now? (0) In one class, all but one student, a philosopher no doubt, placed his mark *close* to zero. The one student, a philosopher no doubt, placed his mark *close* to zero. He stated that it was entirely possible that scientists might invent a pill granting humans such longevity.
J. A box contains slips of paper colored red, white, blue, green, and black. If you reach for a slip, what is the probability that it will be red, white, blue, green, or black? (1) Such examples emphasize the importance of *one* in the field of probability—a certainty in the world of mathematics.
K. A box contains slips of paper marked 1, 2, and 3. What is the probability that a slip drawn will be a 2 or greater? (⅔)
L. What is the probability that the slip of paper you draw will be a 6? (0) The purpose of this type of problem is to reemphasize the notion that an impossible occurrence has a value of zero.
M. What is the probability that a thumbtack, when tossed, will land with the point up? (This problem may lead to the idea that the probability of the occurrence of events of this type depends upon the results obtained in a series of experiments. Thus if one thousand tosses of a thumbtack result in a point-up total of 600, our best estimate of the probability of a tack landing point up will be .6. This means that our next toss of one thousand such thumbtacks should result in an approximately 600 points up.)
N. I am thinking of one of the boys in the class. What is the probability that you can guess his name? (¹⁄₂₀, with twenty boys in the class.)

Other Probability Experiments

1. Put in a box twenty-six slips of paper marked with the letters of the alphabet. Each person in a class writes down a letter. (If the class contains fewer than twenty-six students, have several mark down more than one letter, to make up a total of twenty-six. If more than twenty-six, some may sit this one out.) Select one slip of paper. If a pupil has the letter you select, he raises his hand. Do this experiment ten times. After each selection, pupils may change their letters or keep the same ones. Under normal circumstances one of the twenty-six students should be successful after each of the ten selections. Keep a total of the number of successes. The total number should be not too far from ten. Students will soon observe the relative predictability of events of this nature.

2. Make four folded slips of paper. Three are blank, one has a cross on it. Mix the slips thoroughly and move around the room, asking a pupil to select one. He opens it, tells whether it is blank or not, refolds the paper, and puts it back in the box. Reshuffle thoroughly and continue until all have had a turn. Keep track of the total number of blanks and crosses. (If there are twenty-eight students in the room and things occur normally, about seven students should be successful in obtaining crosses.)

3. Put in a box twenty-six slips of paper marked with the letters of the alphabet. Move around the room, having pupils select a slip of paper, look at it, fold it up again, and return it to the box. After twenty-six times, about how many will draw vowels? (5) (Agree on A, E, I, O, and U as the only vowels.)

Introduction to Fractions

Let us start with a complete square. When we divide the square into two equal parts, each part represents one-half of the square

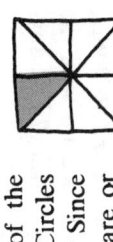

If a square is divided into four equal parts, each part is a representation of one-fourth ($1/4$) of the square. We may also say one-quarter ($1/4$). Circles may also be divided into fourths quite easily. Since four of these quarters make up the entire square or circle, we say that four quarters complete the whole ($4/4 = 1$). Since two of these quarters make up half of the square or circle we say that two quarters are the same as one half ($2/4 = 1/2$). Three of these quarters may be represented as the fraction $3/4$.

If we divide a square into eight equal parts, we may refer to one of these equal parts as one-eighth ($1/8$). Two of these equal parts become $2/8$, which is the same idea as $1/4$. Three of these equal parts may be represented as the fraction $3/8$; four become $4/8$, which is the same idea as $1/2$. Five become the fraction $5/8$; six represent the fraction $6/8$ or $3/4$. Seven become the fraction $7/8$, and eight of the eight parts represent the entire square; that is, $8/8$ is the same as 1.

In each of the following, what fractional part of the diagram is in color? What fractional part of the diagram is spotted? What fractional part is clear?

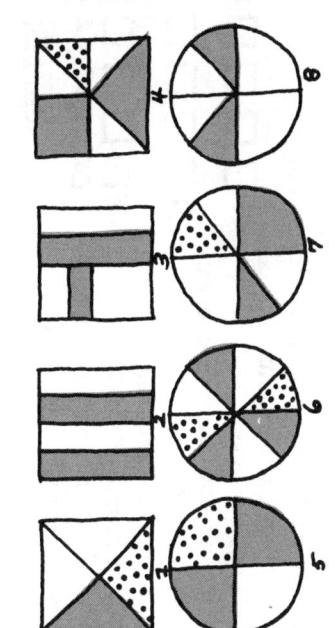

Numbers 9 to 12 may be used to introduce the idea of thirds and sixths. Numbers 13 to 17, which use the regular hexagon, combine the ideas of halves, thirds, and sixths.

($1/2$). In the following drawings, the square has been divided into two equal parts. Each of the parts represents half ($1/2$) of the square. The same idea is then used with circles and triangles.

There are many, many other ways of dividing squares, circles, and triangles into two equal parts, or halves.

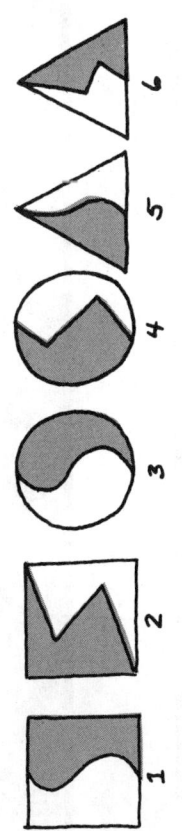

The first four diagrams above have been divided into two equal parts which happen to be congruent—exactly the same size and shape. Diagrams 5 and 6 involve different considerations. Here the parts are equal but not congruent. It would take the same amount of paint to color the left half of each triangle as to color the right half. The idea of halves implies equality, but not necessarily congruency.

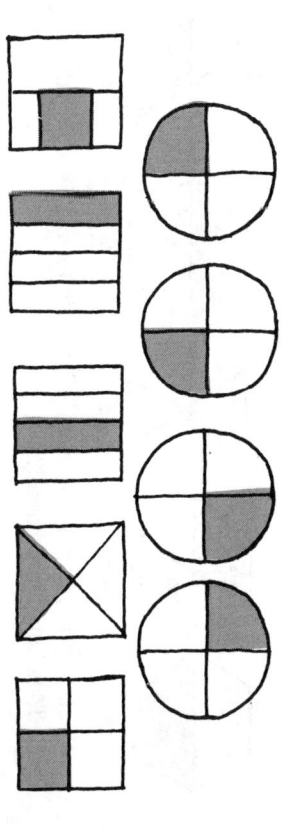

Patterns

11

Using Desks and Seats

With nothing but the desks and seats in the room, many other interesting mathematical problems can be solved. For example, line up the school desks and chairs by columns and rows as in the diagram below. Each child writes down his column number and row number *in that order*. This index number is called a pupil's CR Rating (Column Row Rating)—1; 3; 5, 3; 4, 4; and so on.

Interesting Maneuvers at the Primary Level
1. Stand if your column number is a 2; if it is a 5.
2. Stand if your column number is a 17. (We hope no one stands.)
3. Stand if your row number is even; if it is odd.
4. Stand if your column number is a multiple of 3; multiple of 4.
5. Sit down if the sum of your column and row numbers is odd; all others stand.
6. Sit down if the sum of your column and row numbers is a multiple of 5; all others stand.
7. Stand if your column number is greater than your row number.
8. Stand if your column and row numbers *are* the same; are *not*.

Mathematical Problems Suited to the Upper Grades
1. A prime number is a natural number divisible by only itself and 1. The first six primes are 2, 3, 5, 7, 11, 13. Stand up if the sum of your column and row numbers is prime.
2. Multiply your column number by 4; your row number by 5. Add the products. Stand if the sum is exactly 41; is a multiple of 4; is 50 or over; is less than 50; is even.
3. Stand if the difference between your column and row numbers is a multiple of 3.
4. Multiply your column number by itself, your row number by itself. Stand if the difference of the two is odd.
5. Multiply your column number by 6; your row number by itself. Add the products. Stand if the sum is a multiple of 6.
6. If A equals 1, B equals 2, and so on, translate your CR Rating into its alphabetical equivalent. How many ended up with the same two initials? Where must they have been sitting? Did anyone end up with the initials ER? (We hope not.)

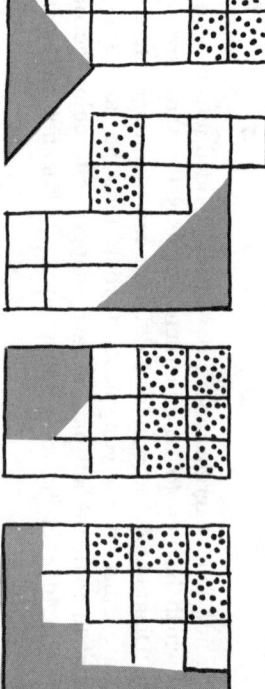

Using the Classroom to Solve Problems

Some of the best teaching aids in mathematics are those so obvious that no one uses them. Consider the ordinary classroom. It is a rectangular solid. This solid has six surfaces or *faces*, parallel by pairs. It has eight corners or *vertices*. There are twelve *edges*; four are lengths, four widths, four heights.
1. What is the largest number of colors we might use to paint the six surfaces of the room?
2. What is the smallest number of colors we might use so that no two adjoining surfaces are colored alike?
3. What is the shortest path from a point on the ceiling to a point on the floor, provided we travel only on surfaces of the room?
4. What is the shortest path from a point on one of the end walls to a point on the opposite wall, under the same conditions?
5. Is there some point within the room itself which is equally distant from the six surfaces of the room?

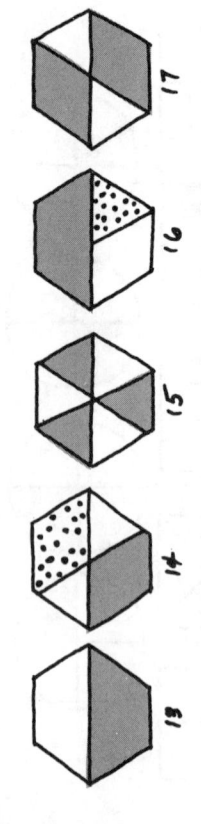

What fractional part of each of the following is colored? Is spotted? Is not colored at all?

Imaginative Ideas: Ranucci's Reservoir

7. Did anyone end up with the initials AC; CA? Where were these two sitting? (This question poses intriguing possibilities with respect to symmetry, mirror images, and so on.)

Making Graphs

Room columns and rows can be easily transformed into graphs. Squares are colored according to a given set of conditions.

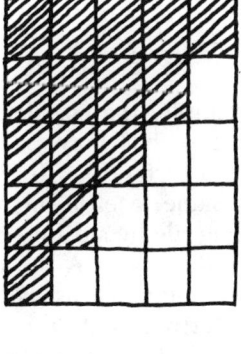

Color those squares where the column number and the row number matches.

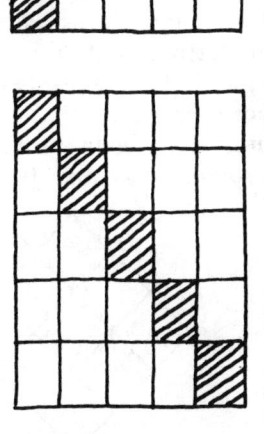

Color those squares where the sum of column and row numbers is less than 5.

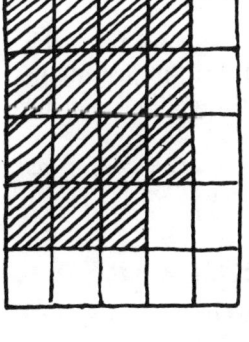

Color those squares where the sum of column and row numbers is greater than 5.

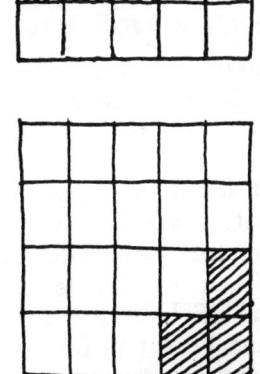

Color those squares where the product of column and row numbers is greater than 5.

Patterns

Mathematics on the Ceiling

by Ernest R. Ranucci

Some of the best teaching aids in mathematics were never planned with this purpose in mind—ceiling tile, for example. Many of the modern classrooms are equipped with it. Some of it is the usual type, ordinary asbestos squares; some of it is the perforated acoustical type. Both may be used to illustrate mathematical concepts.

The big advantage of ceiling tile is its immediacy, and its large scale makes it convenient. It costs nothing and is readily available through a slight bending of the neck muscles. It never needs erasing.

In some cases it is advantageous for each student to have a tile in hand. But, in just about every case to be described, the ceiling may be used for inspiration.

Here are a few of the uses of the ceiling as a mathematics teaching device.

BINOMIAL THEOREM

There was a time, about 10 or 15 years ago, when the Binomial Theorem showed evidences of having outworn its usefulness in the field of algebra. This topic, usually included in intermediate algebra, was taught when time permitted; more often it was not taught. It was looked upon as an interesting trick, but with not much application. This was in the era before probability became so vital a field of study. Now we find some of it being taught in forward-looking elementary programs.

Probability is chock-full of applications of the Binomial Theorem and the ceiling may be used to develop this interesting and useful mathematical maneuver.

Consider the expansions of
$(a + b)^1 = a + b$
$(a + b)^2 = a^2 + 2ab + b^2$
$(a + b)^3 = a^3 + 3a^2b + 3ab^2 + b^3$
$(a + b)^4 = a^4 + 4a^3b + 6a^2b^2 + 4ab^3 + b^4$

Pascal's Triangle predicts these coefficients.

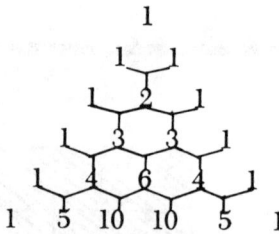

For example, the coefficients for the expansion of $(a + b)^5$ may be obtained from the sixth row (1 5 10 10 5 1).

These same coefficients may be obtained from ceiling tile by a different process. Consider Figure 1. Starting at A the routes from A to B and C are unique. This we indicate as 1,1.

Again starting at A, the routes to D and F are unique. But there are two routes to E. This we indicate as 1,2,1.

Starting at A again in Figure 1, there are unique routes to G and J; triple routes to H

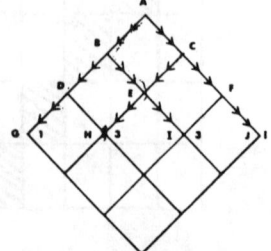

Figure 1

and I. Thus: 1,3,3,1. The student can establish the surface validity of these statements by "reading" the ceiling. Later on he will be able to validate these statements through more orthodox mathematical approaches.

The binomial theorem has all sorts of uses in problems concerned with probability, some of which can be made understandable to junior high school students. For example:

If one coin is tossed, the probability of obtaining a head is ½. The probability of obtaining a tail is ½.

If two coins are tossed, the probability of obtaining two heads is ¼; obtaining one head and one tail, 2/4; two tails, ¼. The different possibilities may be exhaustively listed in tabular form as follows:

Outcome of Coin No. 1	Outcome of Coin No. 2
H	H
H	T
T	H
T	T

The binomial theorem predicts this outcome through the coefficients of the third row of Pascal's triangle, 1,2,1. The denominator "4," which represents total possibilities, can be found by summing the numbers of that third row, or by evaluating the expression 2^n where n represents the number of coins used.

A toss of three coins results in the following possible arrangements: Three heads, two heads and one tail, one head and two tails and three tails. The respective probability for each of these arrangements is: ⅛, ⅜, ⅜, ⅛. It can be seen that the third row of Pascal's triangle or the coefficients of $(a + b)^3$ predicts the numerators of these probability fractions and the sum of that row, or 2^3, provides the denominator. From the table below, which exhausts the possible arrangements, it can be seen that there are three ways of obtaining two heads and a tail and three ways of obtaining two tails and one head.

Outcome of Three Coins

Coins:	1	2	3
	H	H	H
	H	H	T
	H	T	H
	H	T	T
	T	H	H
	T	H	T
	T	T	H
	T	T	T

NEIGHBORS

Figure 2 shows an individual central ceiling tile surrounded by eight adjacent ones.

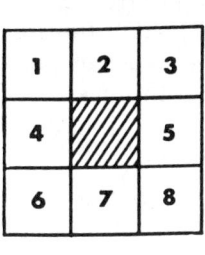
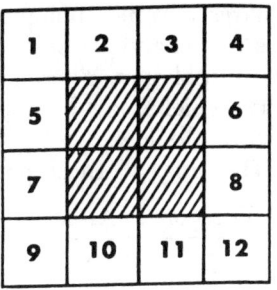

Figure 2   Figure 3

Figure 3 shows that a central cluster of four of these tiles (arranged in a square pattern) are adjacent to 12. We can count on the ceiling that 16 are adjacent to a center cluster of nine. Now the question: Is the number of tiles adjacent to a square cluster of tiles predictable without counting?

Evidently one tile rests on each exposed side of a tile in the original square cluster. And one tile rests on each "exposed" corner. Let the students "play" with these numbers for these examples and see if they can discover that the sum of the number of the "side" tiles and "corner" tiles is the desired predictor. Then let them predict, and check, the results for square clusters of 25, 36, etc., tiles.

Eventually they should read the formula which predicts this sum, $4s + 4$, where s represents the number of tiles or squares on a side of the central square cluster.

An extension of this exercise to three dimensions follows readily from the tile analogy. How many cubes will surround any given cube? Every face of the cube has a cube on it, so that means 6. Every edge of the cube has a cube adjacent to it, so that is 12 more. Every corner of the central cube has a cube adjacent to it, so that is 8 more, or 26 cubes in all surrounding the original cube. This is easily verified by thinking that the central cube and all its neighbors would appear as a large cubical arrangement, 3 cubes along each edge, or $3^3 = 27$ cubes in all. Removing the original central cube means that its "neighbors" must have totaled $27 - 1$, or 26 in all.

The student should try to discover a formula which will predict the number of cubical "neighbors" to a given cubical cluster of cubes. The previous derivation indicates that the formula will be: $6s^2 + 12s + 8$. Here s represents the number of cubes along one edge of the cubical array.

Patterns

AREA OF POLYGONS

Each of the unit squares on a ceiling has an area of one square unit. The vertices of these squares may be thought of as an array of lattice points and are sufficient to predict the area of any polygon. There are many types of polygons which may be formed in this way. See Figure 4 below.

The areas of these polygons are:

Polygon	Area	Polygon	Area
1	1	6	4
2	½	7	3
3	3	8	3
4	3½	9	5
5	1	10	7

The interesting question: Is there some method whereby areas may be computed by counting *only* lattice points?

Note that the polygons vary with respect to these lattice points. Some have lattice points which lie on the perimeter of the polygon ("exterior" points) only. Some have lattice points which lie in the interior as well as lattice points which are exterior. Some are composite in nature.

Let us tabulate the details for the ten polygons of Figure 4 with respect to these two conditions.

Polygon No.	Exterior Points	Interior Points
1	4	0
2	3	0
3	8	0
4	7	1
5	4	0
6	8	1
7	8	0
8	6	1
9	8	2
10	10	3

Areas of polygons which involve exterior points only (Polygons Nos. 1, 2, 3, 5 and 7) may be computed by a formula that evolves from the following reasoning:

Using the unit square as the fundamental structure under consideration (Polygon No. 1 of Figure 4) we note that its vertices consist of four exterior lattice points. But only two of these lattice points are needed to outline the base of this unit square, half as many as the 4 which form its full complement of vertices.

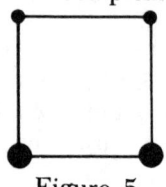

Figure 5

Since *two* points determine only *one* line segment, a base flanked by two end points will correspond to only *one* unit square. The area of a unit square may thus be determined by halving the number of vertex points and diminishing by one ($A = \frac{1}{2} V - 1$). In a similar manner three points outline two segments, four points outline three segments, etc.

This formula may be confirmed by considering Figures 6 and 7.

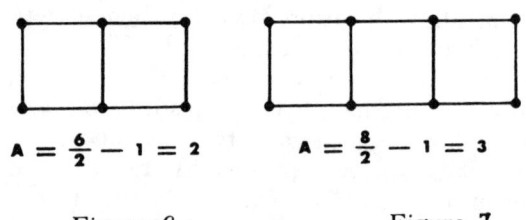

Figure 6   Figure 7

Simple areas which are concerned with exterior points only may be computed from the same formula.

When a vertex is shared (no common edge) as in Figure 8, the formula does not work.

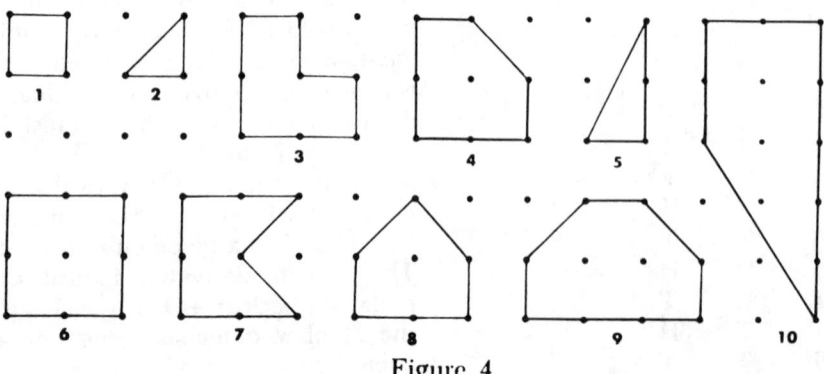

Figure 4

Imaginative Ideas: Ranucci's Reservoir

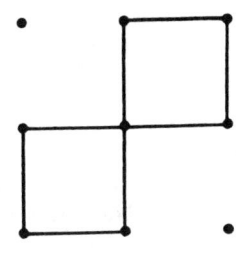

Figure 8

$$A = \frac{7}{2} - 1 = 3\frac{1}{2} - 1 = 2\frac{1}{2}$$

The area of the figure in Figure 8 is, obviously, 2. When vertices are shared in this way, i.e., no common edges, areas must be computed as separate polygons. For example, the area of Figure 9 by direct substitution in the

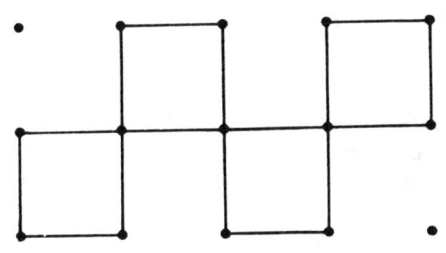

Figure 9

original formula $A = \frac{1}{2}V - 1$ would predict an area of 5½, instead of the obvious true area, 4.

In those cases which involve both interior and exterior points a different procedure must be followed. Consider Figure 10a. Here the

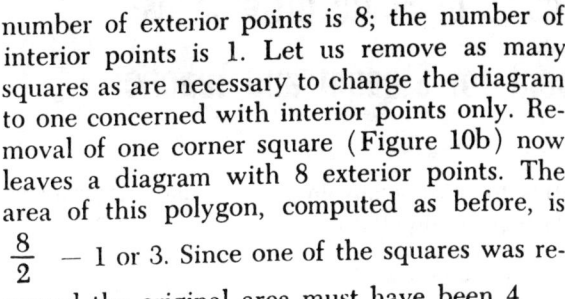

Figure 10

number of exterior points is 8; the number of interior points is 1. Let us remove as many squares as are necessary to change the diagram to one concerned with interior points only. Removal of one corner square (Figure 10b) now leaves a diagram with 8 exterior points. The area of this polygon, computed as before, is $\frac{8}{2} - 1$ or 3. Since one of the squares was removed the original area must have been 4.

Consider Figure 11a. Here the number of

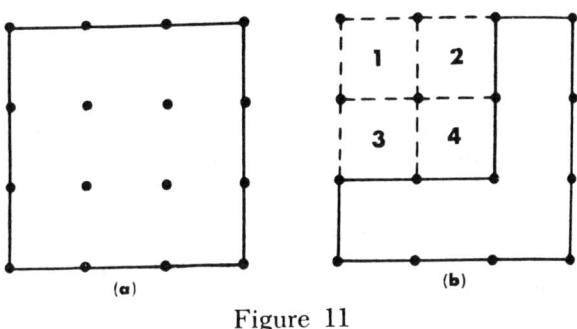

Figure 11

exterior points is 12; the number of interior points, 4. Removal of the squares marked 1 through 4 (Figure 11b) leaves a polygon whose lattice points are all exterior. The area then is $\frac{12}{2} - 1$ or 5. Since four square units were removed, the original area must have been 9.

The interesting fact here is that the number of squares whose removal would result in a polygon with exterior points only is always the same as the number of interior points. Because of this fact, the formula applicable to polygons whose lattice points are both interior and exterior is $\frac{E}{2} - 1 + I$.

Figure 12 below shows a number of polygons whose areas may be computed by traditional means, and then verified by the use of lattice formulas.

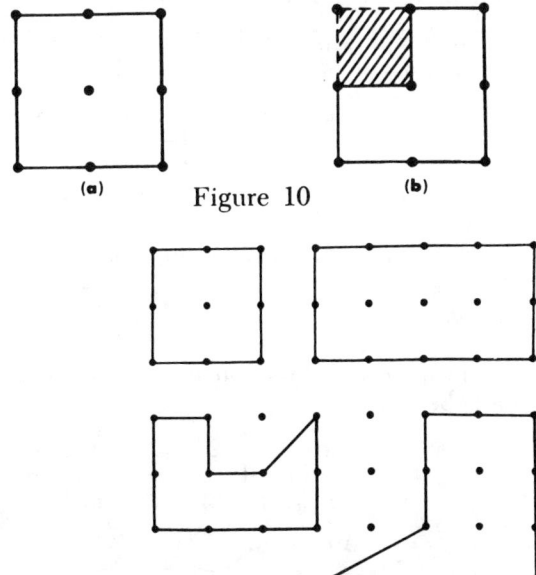

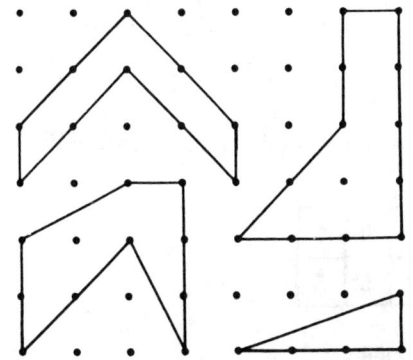

Patterns 17

The Calculus of Finite Differences: Enrichment for Student and Teacher

Opportunities to make many interesting and important discoveries in mathematics arise from observation of simple things around us. In fact, one of the marks of the mathematical or scientific "brain" is a curiosity about anything and everything. Here are some examples of questions which arise in the inquisitive mind:

• When mud dries up and fissures form, is there a typical pattern to these figures or are these cobweb-like networks merely a matter of chance?

• Is there a way of figuring out the shortest route between street intersections in a typical group of city streets?

• If a box is loaded with identical spherical clay pellets and pressure is applied equally from all directions, what do the resulting solids look like?

These examples are essentially nonnumerical in content. They are typical of mathematical problems which may be attacked by experimental means. On the other hand very often we are presented with evidence of a numerical nature and, on the basis of this evidence, we may try to generalize. Some examples:

• When tenpins are set up in their usual pattern (see Fig. 1), we know that if there are 40 such rows of pins, we have 40 pins in the fortieth row. Easy enough, but can we predict the total number of pins in the entire collection of 40 rows?

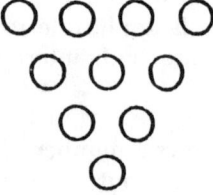

Figure 1

• A diagram such as that in Fig. 2 has 18 separate rectangles in the entire configuration. Using informal notation, they may be recorded as A, B, C, D, E, F, AB, BC, DE, EF, AD, BE, CF, ABC, DEF, ABDE, BCEF, and ABCDEF. Can we predict the total number of rectangles in a diagram with 30 cells on one side, and 20 cells on the other?

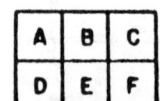

Figure 2

• Two points divide a line segment into six separate segments (see Fig. 3). They are: AB, BC, CD, AC, BD, and AD. But can we predict the total number of segments formed when 1000 separate points are placed between the ends of a line segment?

Figure 3

In each of these cases the initial problem is readily solved by counting; the second is not. ACTUAL COUNTING IS SO UNSATISFACTORY!

We have raised the problems. Now let us turn to the development of a general solution technique that will suit these problems and many others. In particular we shall deal with an important technique taken from the calculus of finite differences. In many cases use of this technique enables us to find the key to mathematical relationships of a numerical nature. The algebra involved is not at all difficult and I have found that high school students well into first year algebra can handle it. The topic serves to draw together many ideas and to give students a practical application for some of the content they have learned.

The Mathematical Background

Algebraic functions of various degrees have certain recognizable family traits. For example:

When x is replaced by 1, 2, 3, 4, and 5, the linear function $y = 6x + 3$ results in the table:

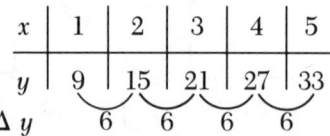

It is evident that the first differences are constant. Note that we have used the symbol Δ (Greek delta) for the phrase "difference in."

The quadratic function $y = x^2 + 6x + 3$, results in the table below:

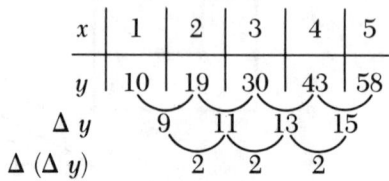

It is evident that while the first differences are *not* constant, the second differences *are*.

The cubic function $y = x^3 + 3x^2 - x + 6$, results in the table below:

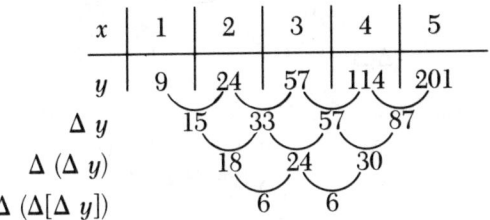

It is evident that, while the first and second differences are *not* constant, the third differences *are*.

The fourth degree function $y = x^4 + x^3 - 2x^2 + x + 1$, results in the table below:

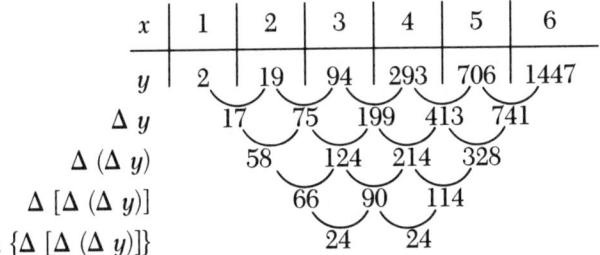

It is evident that, while the first, second, and third differences are *not* constant, the fourth differences *are*.

Let us now examine more thoroughly the *generalized* forms of these functions.

The generalized linear function has the form $y = ax + b$. Substitution of appropriate numbers results in the following table:

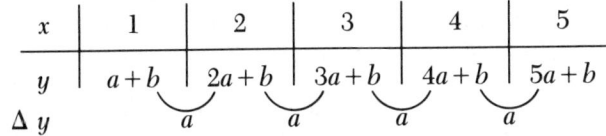

(I) The *constant* difference is a.

(II) The generalized quadratic function has the form: $y = ax^2 + bx + c$. Substitution of appropriate values of x, results in the following table:

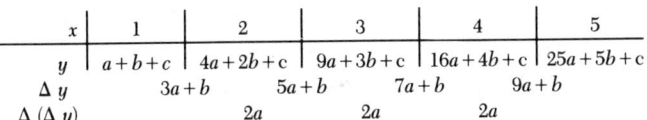

The *constant* second difference here is $2a$.

(III) The generalized cubic function has the form: $ax^3 + bx^2 + cx + d$.

Substitution of appropriate values of x results in the following table:

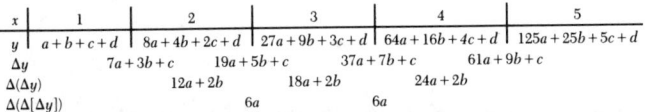

The *constant* third difference is $6a$.

(IV) By similar means the general fourth degree function $y = ax^4 + bx^3 + cx^2 + dx + e$ may be shown to have a *constant* fourth difference, $24a$. In fact, it may be shown that a general function of degree n, will have the nth differences constant.

Applying the Basic Idea

We may now use certain of these functions to answer questions that come up in the course of mathematical research of an elementary nature. Let us first examine two typical problems connected with arithmetic progressions.

1. Find the 50th term of the arithmetic progression, 1, 3, 5, 7, 9, . . .

We let x represent the number of the term and y represent its value. Tabulating this information we have:

x	1	2	3	4	5
y	1	3	5	7	9
Δy		2	2	2	2

We note that the first differences are alike so our equation must be of the form $y = ax + b$. We may now use the information we have about this type of function gathered in (I). Reference to (I) shows that:

From Δy: $a = 2$ From y: $a + b = 1$
$2 + b = 1$
$b = -1$

The function must then be: $y = 2a - 1$ and the 50th term (substitute $x = 50$) must be 99. This may be confirmed by the usual formula for the nth term of an arithmetic progression or for these junior high youngsters by counting to the 50th odd number.

2. Find the sum of fifty terms of the same series: 1, 3, 5, 7, 9, . . .

Again we let x represent the number of the term, but now y represents the sum of terms to this point. Our sequence of sums is 1, 1 + 3 = 4, 1 + 3 + 5 = 9, 1 + 3 + 5 + 7 = 16, etc.

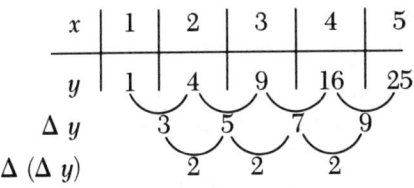

Since the second differences are constant, the function must be quadratic.

Reference to (II) shows that:

From $\Delta(\Delta y)$: $2a = 2$ From Δy: $3a + b = 3$
$a = 1$ $3 + b = 3$
                                         $b = 0$

From y: $a + b + c = 1$
$1 + 0 + c = 1$
$c = 0$

The function is: $y = x^2$ and the sum of 50 terms of the series ($x = 50$) is 2500.

Patterns

This may be confirmed through use of the standard formula for the sum of n terms of an arithmetic progression, $\frac{n}{2}(a+l)$, or by noticing that each value of y in the table is the square of the corresponding x-value.

3. If there are 500 points on a plane, no three of which are collinear, how many possible segments can be drawn?

The actual experimentation of Fig. 4 results in the following table:

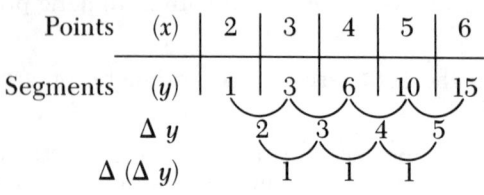

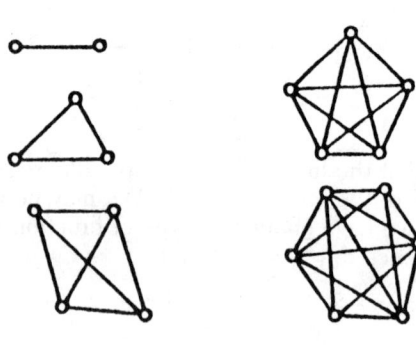

Figure 4

Since the second differences are constant, the function must be quadratic. Application of formula (II) results in the function:

$$y = \frac{x(x-1)}{2}$$

From $\Delta(\Delta y)$: $2a = 1$
$a = \frac{1}{2}$

From Δy: $5a + b = 2$
$\frac{5}{2} + b = 2$
$b = -\frac{1}{2}$

From y: $4a + 2b + c = 1$
$2 - 1 + c = 1$
$c = 0$

There must be 124,750 separate segments formed when 500 separate points are joined in pairs.

An interesting side question occurs here: What changes would have to be made if the 500 points were situated in three-space, not on the same plane? Answer: None. Two points still determine one segment.

4. How many separate triangles would be formed under the conditions of problem 3, provided that only the *original* points might be considered as possible vertices of these triangles? Direct experimentation with actual counting as in Fig. 4 results in the following table:

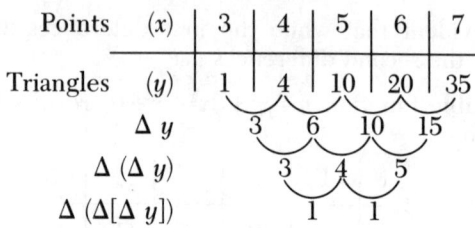

Since the third differences are constant, the function must be cubic. Application of formula (III) results in the function: $\frac{x^3}{6} - \frac{x^2}{2} + \frac{x}{3} = \frac{x(x-1)(x-2)}{6}$

There must be 20,708,500 separate triangles formed when 500 points, no three of which are collinear, are joined in all possible combinations.

The sidelight corresponding to that of question 3 is:

What changes would have to be made if the 500 points were not coplanar? Again the answer is none. Three points still determine only one triangle.

What We Have Accomplished

Here we have developed a powerful mathematical tool, well within the ability of algebra students. In teaching this procedure excellent opportunities are offered for equation development and solution, experimentation, manipulation of algebraic expressions and computations, all exercises students need. The alert student will store away this technique for the great many future uses which will be open to him.

One final warning. Beware that you or they should think that this method is all-powerful as we have stated it. Many further developments must be added when students are better prepared to understand them. One way to show the shortcomings of the technique is to take a sequence like the following (the number of moves y required to complete the game Tower of Hanoi when x discs are used):

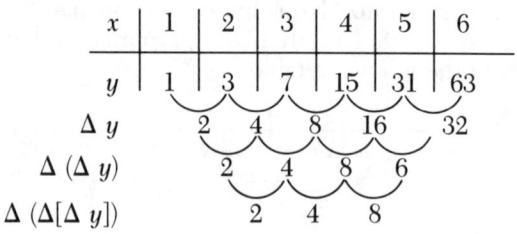

It becomes obvious that these differences will continue to repeat forever. The formula is, however, although not of polynomial form, a simple one:

$$y = 2^x - 1$$

Ernest R. Ranucci, Coordinator of Mathematics, American Elementary and High School, Caixa Postal 7432, Sao Paulo, Brazil.

Discovery in mathematics

ERNEST R. RANUCCI

International School Services, Washington, D.C.

Dr. Ranucci is presently under contract with International School Services for the academic year, doing work in mathematics education in eight countries in South America.

Many interesting facts may be discovered by elementary school students who observe things right around them. For example, the pins in a bowling alley are set up in this way: There is one pin in

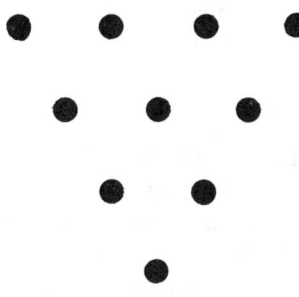

the first row. There is a total of three pins in the first two rows; a total of six pins in the first three rows; a total of ten pins in the first four rows. Were there *five* rows, the total would have been fifteen.

The series represented by these numbers, 1, 3, 6, 10, 15, etc., comes up time and time again in mathematics. Children seem to enjoy further investigation of this series. For example:

Two points A and B determine exactly one segment, $\overleftrightarrow{AB}$.

Three points in a row determine three segments. These are $\overleftrightarrow{AC}$, $\overleftrightarrow{CB}$, and $\overleftrightarrow{AB}$.

Four points in a row (the proper word is *collinear*) determine six segments: $\overleftrightarrow{AC}$, $\overleftrightarrow{CD}$, $\overleftrightarrow{DB}$, $\overleftrightarrow{AD}$, $\overleftrightarrow{CB}$, and $\overleftrightarrow{AB}$.

Five collinear points determine ten segments: $\overleftrightarrow{AC}$, $\overleftrightarrow{CD}$, $\overleftrightarrow{DE}$, $\overleftrightarrow{EB}$, $\overleftrightarrow{AD}$, $\overleftrightarrow{CE}$, $\overleftrightarrow{DB}$, $\overleftrightarrow{AE}$, $\overleftrightarrow{CB}$, and $\overleftrightarrow{AB}$.

Six collinear points determine fifteen segments: $\overleftrightarrow{AC}$, $\overleftrightarrow{CD}$, $\overleftrightarrow{DE}$, $\overleftrightarrow{EF}$, $\overleftrightarrow{FB}$, $\overleftrightarrow{AD}$, $\overleftrightarrow{CE}$, $\overleftrightarrow{DF}$, $\overleftrightarrow{EB}$, $\overleftrightarrow{AE}$, $\overleftrightarrow{CF}$, $\overleftrightarrow{DB}$, $\overleftrightarrow{AF}$, $\overleftrightarrow{CB}$, and $\overleftrightarrow{AB}$.

The following table summarizes these discoveries. From such a table we may predict future values without too much actual counting.

Number of points	Number of segments	Differences
2	1	
		2
3	3	
		3
4	6	
		4
5	10	
		5
6	15	
		6
7	21?	

It looks as though seven points should give us 21 segments. Actual counting will give us the following segments:

Patterns

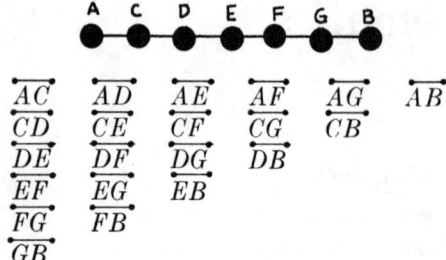

This last table looks suspiciously like the sum $6+5+4+3+2+1$; this means that a total of 8 points should yield:

$7+6+5+4+3+2+1$ or twenty-eight segments. It does. It is also exactly the way pins in a bowling alley are set up. Its structure is said to be *isomorphic* to that of the bowling alley problem.

Investigating squares—1

A square 1×1 contains exactly one square shape.

(1)

A square 2×2 contains, in all, five square shapes: the original and four small squares.

$(1) + (4) = 5$

A square 3×3 contains, in all, fourteen square shapes. There is one square 3×3, four squares 2×2, and nine squares 1×1.

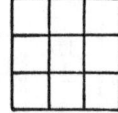

$(1) + (4) + (9) = 14$

A square 4×4 contains thirty square shapes. There is one square 4×4; there are four squares 3×3; there are nine squares 2×2, and sixteen squares 1×1.

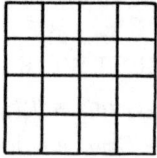

$(1) + (4) + (9) + (16) = 30$

It looks as though a square 5×5 should contain 55 square shapes.

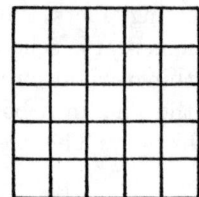

$(1) + (4) + (9) + (16) + (25) = 55$

Find out if this is so.

Investigating squares—2

Let us take the squares used in the preceding exercise and investigate the total number of *rectangles* contained in each figure. Remember that every square is also a rectangle, which means that we may count every square previously found *and* any new rectangles discovered.

The square 1×1 still contains only one rectangle.

The square 2×2 contains nine rectangles:

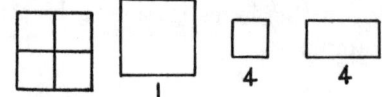

The square 3×3 contains thirty-six rectangles:

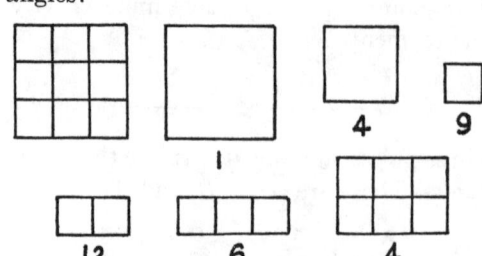

The square 4×4 contains one hundred rectangles:

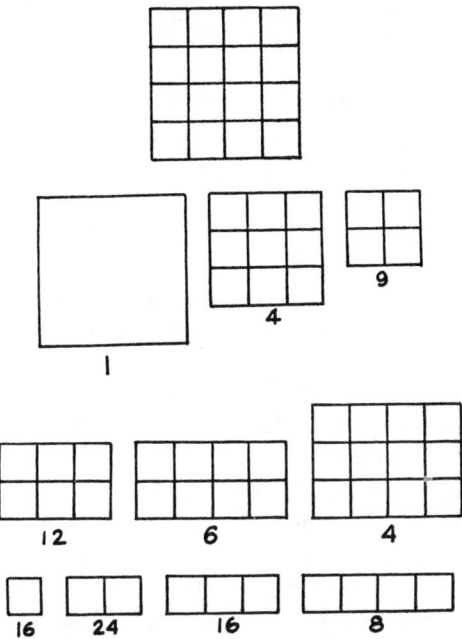

Actual counting is getting difficult; perhaps we can *predict* the total number of rectangles for a square 5×5.

For the square	Total number of rectangles	
1×1	1 = 1×1	*But these last numbers were the same numbers we had in the bowling alley problem....*
2×2	9 = 3×3	
3×3	36 = 6×6	
4×4	100 = 10×10	

It looks as though a square 5×5 should yield 15×15 or 225 separate rectangles. If you feel ambitious, you might want to check this amazing total in Figure 1 at the bottom of the next column.

Permutations discovered through circles

Mark a point on a circle. Call it Point A.

Mark two points on a circle. Connect them. Call the segment thus formed $\overline{AB}$ or $\overline{BA}$.

Three points on a circle A, B, and C, may form the vertices of a triangle ABC. There are six different names for this triangle. They are: ABC, ACB, BAC, BCA, CAB, CBA. We call these six different orders *permutations* of the letters A, B, and C.

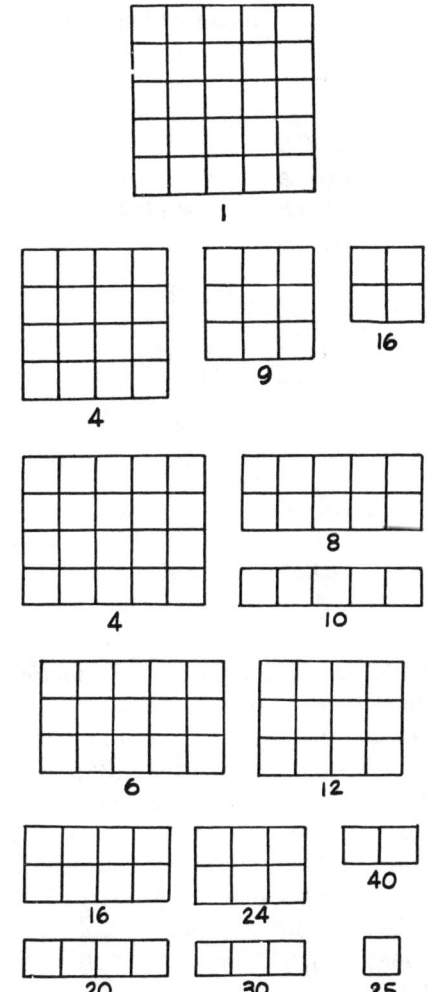

Figure 1

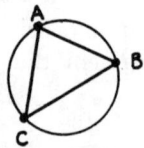

Four points on a circle may be joined in various orders. In listing the various possibilities, we start at any vertex, trace the figure completely, then return to the original vertex.

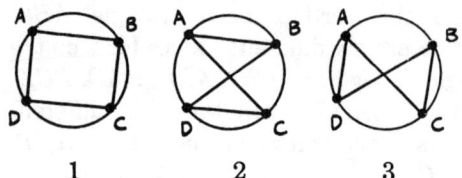

Each of the three diagrams has eight distinct possibilities. They are:

Diagram 1
ABCD	DCBA
BCDA	ADCB
CDAB	BADC
DABC	CBAD

Diagram 2
ABDC	CDBA
BDCA	ACDB
DCAB	BACD
CABD	DBAC

Diagram 3
ACBD	DBCA
CBDA	ADBC
BDAC	CADB
DACB	BCAD

These 24 possibilities are the *permutations* of the letters A, B, C, and D. We may arrive at this number another way, through the use of the "tree" (Fig. 2).

The easiest way to predict the number of permutations does not even require counting, only multiplication. Since the first letter may be one of 4, the second one of 3, the third one of 2 and the last one of 1, we may obtain 24 by simply multiplying 4×3×2×1.

The same experiment performed with five points on a circle yields twelve possible diagrams (Fig. 3). An analysis of just the first of the twelve diagrams shows the following possibilities for naming the pentagon, ten in all.

ABCDE	EDCBA
BCDEA	AEDCB
CDEAB	BAEDC
DEABC	CBAED
EABCD	DCBAE

Since each of the other diagrams will probably yield ten more orders, we may conclude that there are 120 different permutations of the letters A, B, C, D, and E. This is predictable as the product of 5×4×3×2×1.

The exercises discussed in this article are intended for students in Grades 5, 6, 7, or 8. Surprisingly enough, youngsters in grades lower than the fifth sometimes discover much of this material by themselves. Since the exercises involve only simple counting and no complicated arithmetic calculations, young children are not handicapped by their inexperience in arithmetic.

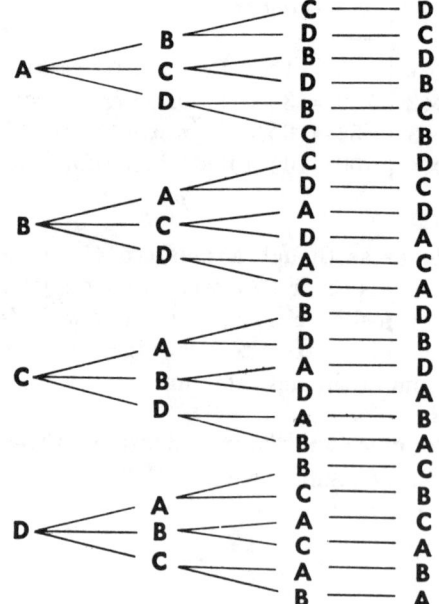

Figure 2

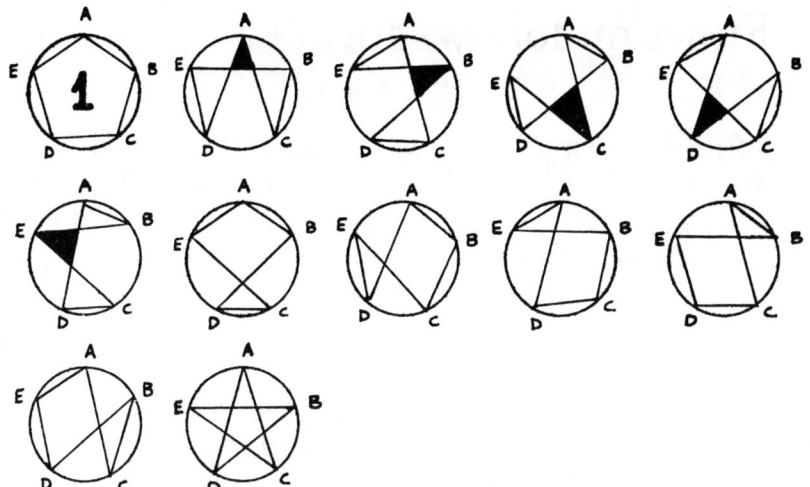

Figure 3

Patterns

Function follows form

ERNEST R. RANUCCI *State University of New York, Albany, New York*

Dr. Ranucci is professor of mathematics education at State University of New York.

The doctrine of the industrial designer seems to be the maxim: Form follows function. This means, I suppose, that the shape of a teaspoon evolved from the cup in which it is customarily used; the Coca-Cola bottle is shaped to fit the hand; the shape of a comb emerged from its use in straightening hair. Perhaps the teacher of arithmetic, and mathematics in general, ought to pay some attention to this advice in an altered form by following the injunction: Function follows form. What I mean by this somewhat flippant alteration of the phrase is that many arithmetic discoveries stem from observation of *form*. Some students function best in arithmetic when their teachers have stressed form in introducing the world of number. Experiments of Piaget and others indicate that the young child is conversant with form much earlier than he is with the world of number. The pre-nursery child can sort a pile of assorted blocks fairly efficiently. All the disks of one size he piles into one group. (He *doesn't* call this a set; the teacher does.) All square blocks the same size go together. The same thing goes for the triangles. Sometimes he places similar forms in piles of graduated size. One of the most popular toys with the young set (excuse the expression) is the set of graduated disks known to the mathematics world as the Tower of Hanoi. My two daughters played for hours with a mailbox toy and its king-sized beads of various shapes. They learned quite easily that the cubical block could be pushed through the square opening; the triangular prism fitted through the triangular opening. They knew intuitively that square pegs just don't fit in round holes. In all of these observations it was *form* which came first; number, if it was involved at all, entered via the back door.

The teacher of arithmetic who consciously uses the world of form sometimes adds a dimension to the world of arithmetic not normally found. Geometry and arithmetic go hand in hand; here I am stressing the *form* aspects of the term "geometry." The Greeks, who were essentially geometers, approached the world of number through form. Many of their discoveries in the field of number came through observation of physical properties of these numbers. "Figurate numbers" was but one of these interests (Figs. 1–3). Many of the students in arithmetic are unaware of the intimate interrelationship between the geometry of form and the number world.

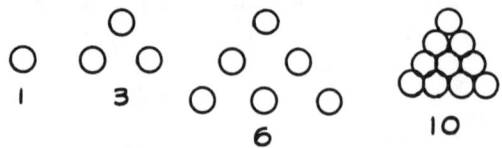

Figure 1. Triangular numbers

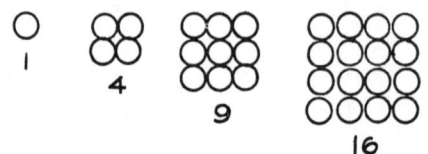

Figure 2. Square numbers

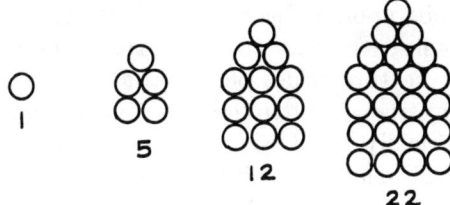

Figure 3. Pentagonal numbers

Another technique which may be used to further explore the function of form in the teaching of arithmetic needs some preliminary explanation. An infinite line, straight or not, *separates* the plane on which it is drawn (Fig. 4). This means, essentially,

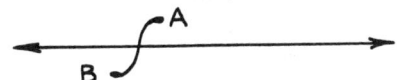

Figure 4

that it is impossible to connect A with B, without intersecting the line somewhere. A circle, for example,

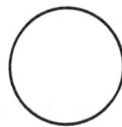

separates the plane on which it is drawn; the inside of the circle could be colored red, the outside white. A circle with a gap does *not* separate the plane on which it is drawn.

If the inside were to be colored, the outside would merely be a continuation of this color.

A series of experiments like the following will help to explain the geometric significance of the terms discussed above. Have the students prepare three columns as in Figure 5. The first, R, will be used to in-

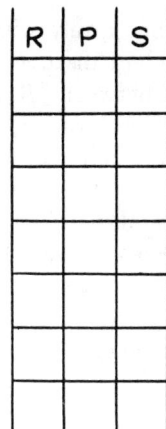

Figure 5

dicate the number of regions in a particular drawing. Column P will be used for the number of points or vertices in general, and column S is reserved for segments. As the teacher draws on the board, the child brings his statistics up to date, row by row (Fig. 6).

	R	P	S
1.	1	1	0
2.	2	1	1
3.	2	2	2
4.	2	3	3
5.	3	3	4
6.	3	4	5
7.	4	4	6

Figure 6

1. A point on a plane lies in a region, the plane.
2. The regions are inside and outside. The circle is counted as one segment like a snake biting its tail.
3. No change in the number of regions. Two points now, two regions.
4. No new regions have been added; the new segment does not separate the plane.
5. A new region has been added.
6. The addition of a new point does not affect the number of regions.
7. The drawing now has four distinct regions.

Sometimes, but rarely without hints, I have had a student discover the relationship existing between R, P, and S: $R+P=S+2$. This formula, attributed to Leonhard Euler, but known before his time, has important adaptations in the world of three dimensions. For simple solids $F+V=E+2$. Here F represents the number of faces of a solid, V represents the number of vertices, and E represents the number of edges. In a cube, for example, $F=6$, $V=8$, and $E=12$.

The regular hexagon, familiar to most students through elementary compass manipulation, offers a wealth of possible imaginative exercises. Here, again, the form world and the number world go hand in hand. Divide a regular hexagon into the following number of congruent parts: 2, 3, 4, 6, 8, 9. In cases where this can be accomplished in more than one way, see how many solutions you can find. For purposes of simplicity let us restrict ourselves to just those solutions which use straight line segments (Fig. 7).

Most of these exercises with the regular hexagon (tessellations) depend upon the fact that the hexagon can be subdivided into twenty-four congruent equilateral triangles (see Fig. 7). This, in turn, depends upon a theorem which states that the line segments which join the midpoints of the three sides of any triangle divide that triangle into four congruent triangles.

Finally, just for fun, let's take a difficult problem and try to explain its solution to a group of intelligent students in the sixth grade. Keep your eye on form and geometry in its explanation: At a certain party, guests shake hands with whomever they please; not everyone shakes hands with everyone else, necessarily. (This is a frightful breach of etiquette in the South American countries where I taught last year: Brazil, Uruguay, Argentina, Chile, Peru, Ecuador, Bolivia, and Paraguay. *Everyone* shakes hands with *everyone* else, both upon coming to a party and upon leaving; I like the custom.) The problem: Show that the number of guests who have shaken hands an odd number of times must always be even (go back and read this again).

What does this have to do with the world of form? Only this: Every handshake may be represented by counting the two endpoints of a segment as a physical representation of the nature of handshaking. For example, the basic nature of segment *AB* is isomorphic (same structure)

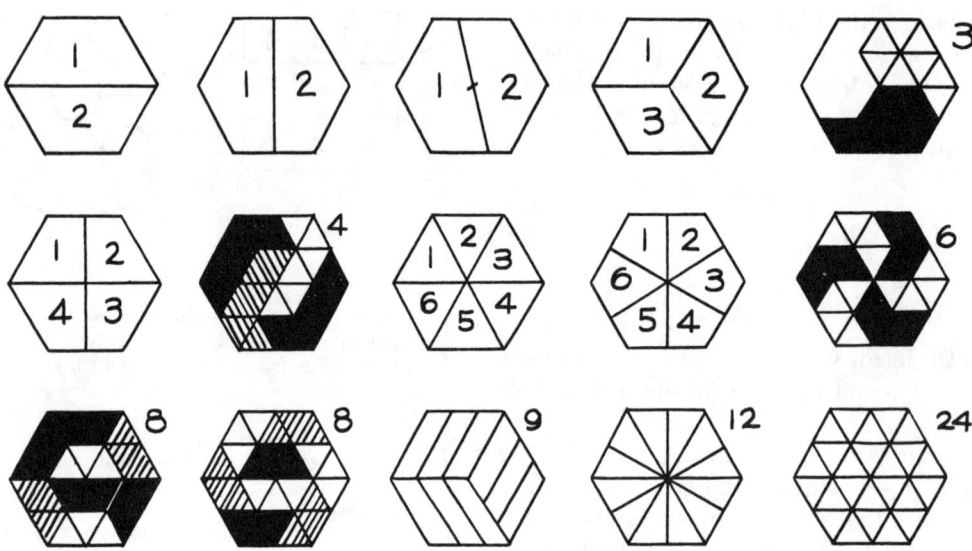

Figure 7

to the situation at a dull party with only two guests. We may place a 1 at both ends, A and B, to indicate that one segment emerges from that vertex (Fig. 8).

Figure 8

Another way of saying the same thing is that a segment has two endpoints. When C joins the party, he may shake hands with both of the others, or snub A, or B. He may also be a misanthrope and wish to shake hands with neither of the others, but we'll make him shake hands with at least one of the others. Don't forget that it takes two to shake hands. The graphs in Figure 9 indicate these possibilities.

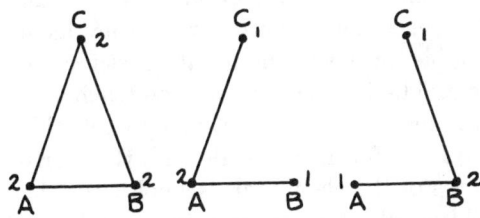

Figure 9

Figure 10 illustrates some possibilities with a party of four, and Figure 11 shows what might happen with a convivial group of five where everyone knows everyone else. "Accidental" intersections don't count. We permit involvement of only those vertices present in the original situation. Figure 12 shows some other theoretical situations.

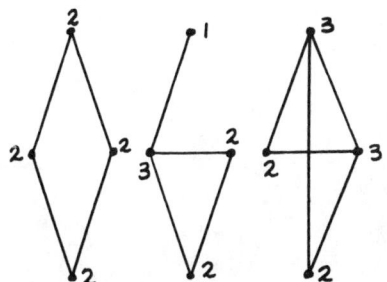

Figure 10

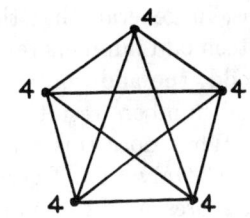

Figure 11

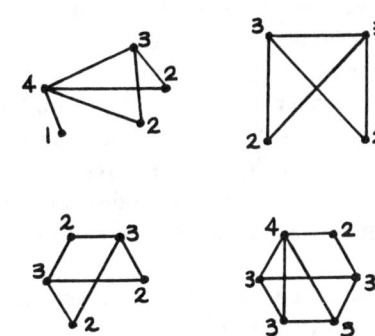

Figure 12

An analysis of the nature of the vertices encountered in all of the previous graphs reveals the following information:

Where even vertices occur, the number of such vertices may be odd or even. In no case where odd vertices occur is the number of such vertices odd.

With special reference to the odd vertices, was this a coincidence? We can't say yet. We do know that the addition table with respect to the addition of odd and even numbers looks like Figure 13. The multiplication table looks like Figure 14.

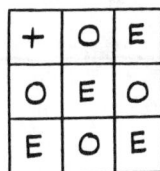

Figure 13

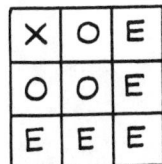

Figure 14

This means, in general, that the addition of seventeen odd numbers results in a sum that is odd; the addition of an even number of odd numbers results in a sum that is even. What does this have to do with our party? Only this: Every handshake may be represented by counting the two endpoints of a segment. If *all* such handshakes are considered, the number of endpoints involved must be an even number. After all, if an even number is multiplied by either an odd or even number, the product must be even. If the character of each of the nodes is considered (odd or even), the number of odd nodes must be even. This is the only way we can obtain an even sum.

I doubt that this rather complex problem is worth the trouble in a sixth grade *or* a seventh grade *or* an eighth grade. This, however, is not the point. What I am trying to say is that the geometrical representation (form) provides the "crutch" to solutions of a rather tricky situation. I have explained this particular problem, with some success, by using a board and nails as a teaching aid. Five nails are driven in lightly as vertices of a pentagon. Pairs of plastic rings are attached to each other via elastic thread. Each of the pairs of rings is colored differently. The explanation goes something like this: A shakes hands with B. (Stretch one of the elastic bands, one ring on A, the other on B.) B shakes hands with C. C shakes hands with D. B shakes hands with E, then B with D.

The result is a pentagon, partially complete, with one ring at A, four at B, two at C, two at D, and one at E (Fig. 15). The

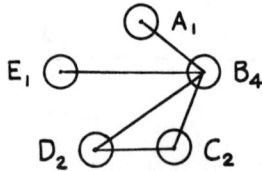

Figure 15

fact that *color* is present here seems to reinforce the notion that even and odd factors are vital to the solution. The essence of the handshake problem is that the sum of the local nodes is twice the number of segments present in the diagram.

To summarize: Function follows form. Children function better when they have tangible or visual material to stimulate them. The implications for the teacher of arithmetic are these. Don't forget this world of form, especially in the early grades of the elementary school. Give the children all the opportunity possible for investigation of the physical world, the world of size, form, shape. Have them cut out patterns, stack blocks of wood, count tin cans, cut potatoes into interesting forms, etc. Realize that the world of mathematics is a complex stew of many ingredients; let the student taste as many of these as possible.

Part 2

Mathematics in the World

Ernest Ranucci was curious about everything that seemed related to mathematics, and perhaps more than many of us, he saw mathematics in the world around him. The two articles selected for this section are examples of this curiosity. The first, "Music in the Marshall Islands," from *School Science and Mathematics,* is a fascinating story of one aspect of the Ranuccis' Peace Corps experience. The musicians among you will know that mathematics is intricately interwoven with music, and your students in band or chorus will appreciate this intermingling of music, mathematics, and international culture.

The second article, "Of Shoes—and Ships—and Sealing Wax— of Barber Poles and Things," from the *Mathematics Teacher,* is a delightful mix of the patterns of mathematics and their real-world counterparts. It is also a portrait of Ranucci, who believed that life was an experience to be enjoyed and that mathematics class was part of that life.

Music in the Marshall Islands

Ernest R. Ranucci
Professor of Mathematics, State University of New York,
Albany, New York

My wife and I were recently involved in the training of teachers and Peace Corps Volunteers in the Marshall Islands. These are twenty-nine in number, with a total area of about 70 square miles. The Marshall Islands, a sub-group of the Pacific Islands, lie about 2,500 miles southwest of Hawaii.

The first Sunday we were in the Marshall Islands—we were on Majuro—we attended the Catholic services. There was no organ, no harmonium, no musical instrument of any type, yet passably good, four-part, unaccompanied harmony was being sung by the Marshallese parishioners. There were hymnals available and a leader started each of the hymns. I picked up one of the hymn books expecting to find European type music notation but found, instead, this:

63. Idem kuelen jar.

```
4/4 ‖ 5  3̄4̄  5   1̇  ·  7  ·  | 4  2̄3̄  4   6   5  ·   ·  '
     3   1   3   4  ·  4  ·  | 7̇  7̇·  7̇   7̇   1  ·   ·  '
  ♩  5   5   5   5  ·  5  ·  | 2  4̄3̄  2   4   3  ·   ·  '
     1  1̄2̄  3   1  ·  2  ·  | 5  ·   5̇   5̇·  1  ·   ·  '

  1. Ñe    e    wa-lok   mä-ram        im    e    jer-kan  ran,

     5  3̄4̄  5   3̇  ·  3̇  ·  | '  1̇   3   7   6  ·   5  '
     3   1   5   5  ·  5  ·  | '  3   1   2   1  ·  7̇  '
     5   5   5̇   1̇  ·  7  ·  | '  6   5   3   ♯4̄  ·  5  '
     1  1̄2̄  3   6  ·  2  ·  | '  2   2   2   2  ·  5̇· '

     Ñe    e    dulem   ja-ko,           i ki-bi-liñ    al:

     5  2̇   6   7   6  ·  5  | 5   2   3   4   3  ·   ·  '
     4   4  7̇·  7̇   1  ·  1  | 7̇·  7̇·  7̇   1  ·   ·  '
     7   7   4   3  ·  3  ·  | 4   5   5   5   5  ·   ·  '
     5̇·  5̇·  5̇   5̇  ·  5̇  ·  | 5   4   3   2   1  ·   ·  '

     Ro rej tä-mak  A - nij,       le-ke An in  nan.

     3  5̇   1   7  ·  6  ·  | 2̇   1   3   ♯4̄   5  ·   ·  '
     2   2   2   2  ·  2  ·  | 3   1   1   7̇·  7̇  ·   ·  '
     ♯5̄  3   ♯4̄  ♯5̄  ·  6  ·  | 6   6   5   6   5  ·   4  '
     3   3   3   3  ·  6  ·  | 1   1   2   2   5̇·  ·   ·  '

     Bue ren wuijdak  Ā - dan          r'-idem kuelen    jar.

     5  3̄4̄  5   1̇  ·  7  ·  | 6   6   2̇   1̇   7  ·   ·  '
     1   1   5   4  ·  4  ·  | ♯4̄   2   5   2   2  ·   ·  '
     3   5   5̇   1̇  ·  1̇  ·  | i ♯4̄   3   ♯4̄   5  ·   ·  '
     1  1̄2̄  3   3  ·  4  ·  | 2   2   2   2   5̇·  ·   4  '

     Je-sus     a-muji  I - roj           kin Am io-kue ir,

     i̇   ·  1̇   2̇   1̇  ·  6  | 5   5   4   2  ·   1   '
     5   ·  5̇   5   5  ·  4  | 3   3   2   7̇  ·   1   '
     i̇   ·  1̇   ♯4̄   6  ·  1̇  | i̇   5   6   4  ·   3   '
     3   ·  3̄·  3   4  ·  4  | 5   5   5̇·  5̇· ·   1   '

     Kuon kä-mä-nän  bur - ueir    ñan air io-kue   Euk!
```

My first thought was that the Marshallese were doing their homework in arithmetic—but on Sunday? . . . and in church? As you have probably guessed, this music notation is an arithmetized version of standard music notation. On my return to the United States, after three months of service, some research disclosed that the music is part of the Justine Ward system. This type of training in music was well-known in the United States between 1920 and 1930. The Ward system had evidently been introduced to the Marshall Islands by missionaries at some time in the past. It is no longer an important part of the music education in the United States. Although it is remarkably easy to learn, it is somewhat unsophisticated and ill-suited to the complexities of contemporary music. It has, however, a lot of merit as an introduction to sight-reading. It provides a bridge between the world of arithmetic and the world of sound.

The basic scale in the Ward system is numbered:

1	2	3	4	5	6	7	1̇
do	re	mi	fa	sol	la	ti	do

The octave above is indicated:

$\dot{1}\quad\dot{2}\quad\dot{3}\quad\dot{4}\quad\dot{5}\quad\dot{6}\quad\dot{7}$

The octave below is indicated:

$\underset{.}{1}\quad\underset{.}{2}\quad\underset{.}{3}\quad\underset{.}{4}\quad\underset{.}{5}\quad\underset{.}{6}\quad\underset{.}{7}$

A slash from northeast to southwest indicates a sharp or, in general, a half-step higher than the note shown.

Thus: ⟋ means F♯ (in the key of C)

A slash from northwest to southeast indicates a flat or, in general, a half-step lower than the indicated note.

Thus: ⟍ means B♭ (in the key of C)

Since the beginning key is always indicated, and since there is no need to conform to standard written music, there is no need for more sophisticated mechanics. A 3 always means *mi*, a third above the tonic (*do*) of the scale. A 5 always means a fifth above (*sol*). All intervals are relative.

The hymn: *Idem kuelen jar*, illustrated above, is written in 4/4 time. The ⟋ means that the basic key is C♯. It has been translated below into standard musical notation. The key has been shifted to C to avoid the complexities inherent to the key of C♯.

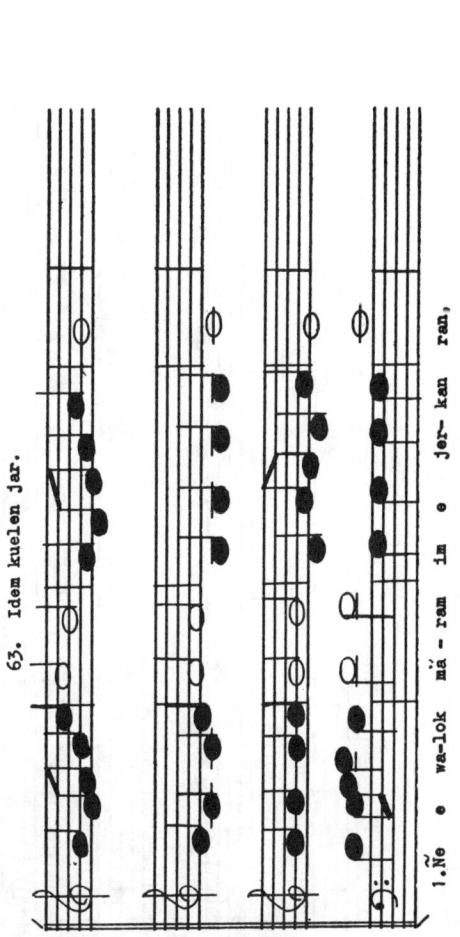

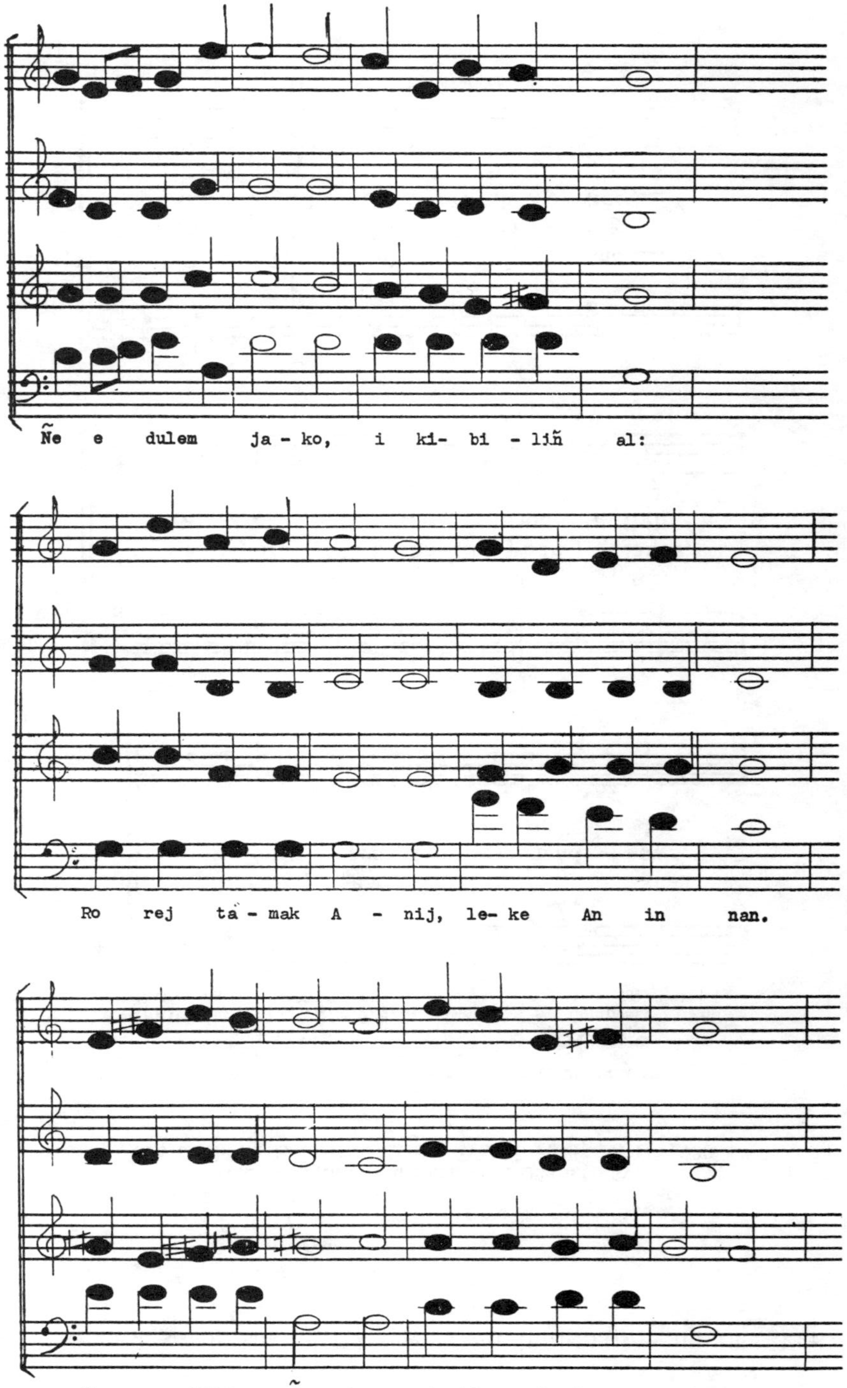

Mathematics in the World

33

The best single source I know of for information on the Ward system is:

Music—First Year
Second Year
Third Year
Catholic Education Series
Ward, Justine and Perkins, Elizabeth W.
The Catholic Education Press, Washington, D. C. 1920

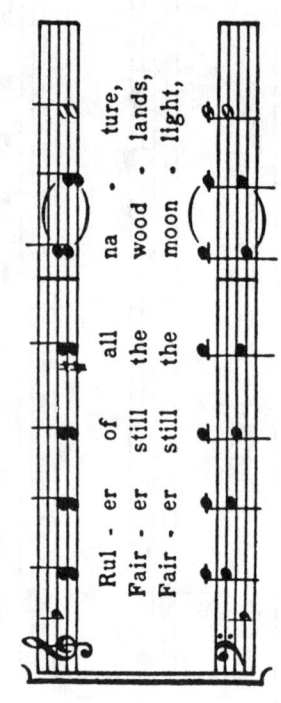
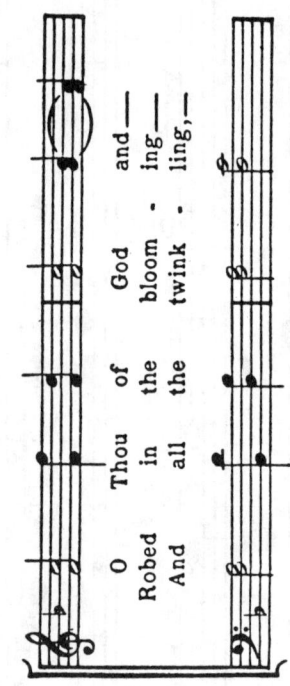

A study of the Ward system is recommended especially for those whose interest is unaccompanied music. The hymn "Fairest Lord Jesus" is included below in both standard music notation and in the Ward system. It is excellent for practice purposes since it includes so few accidentals.

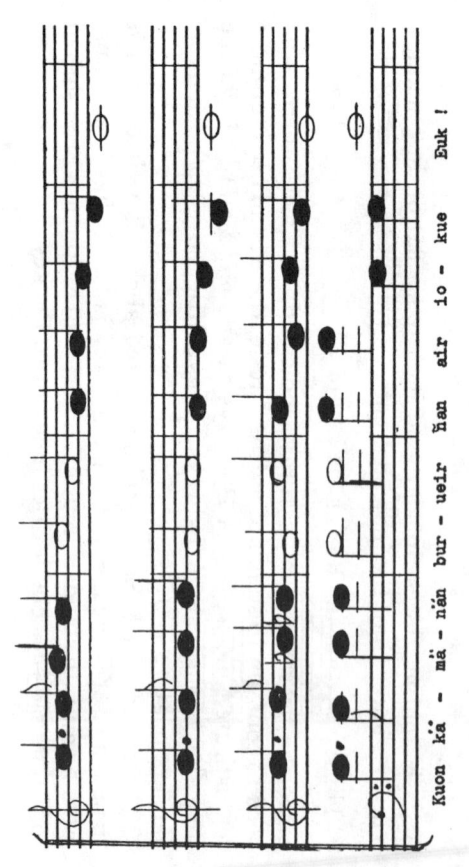

Imaginative Ideas: Ranucci's Reservoir

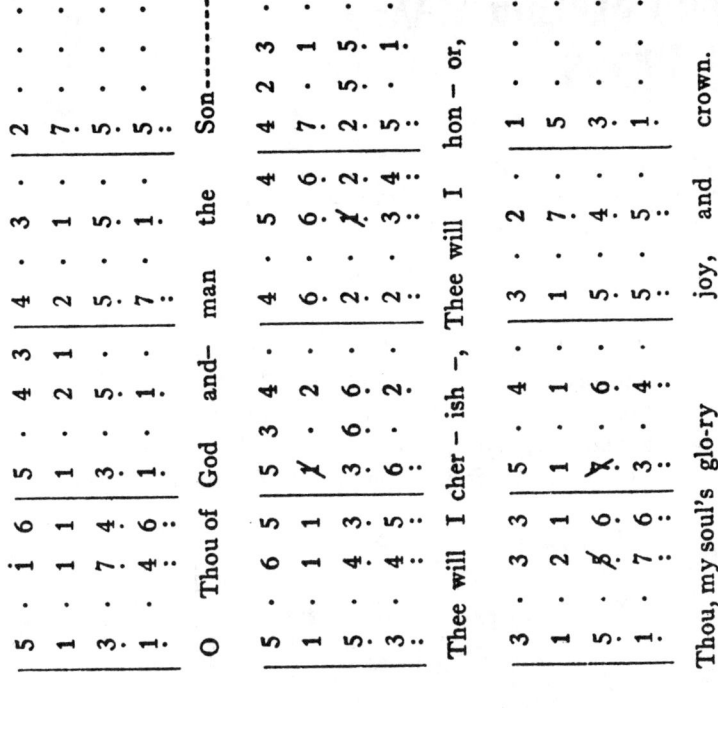

Fairest Lord Jesus

Mathematics in the World

OF SHOES—AND SHIPS—AND SEALING WAX—
OF BARBER POLES AND THINGS

*A collection of whimsical and practical uses of helical designs—
complete with easy-to-follow rules for construction that will
delight mathophiles of all ages.*

By **ERNEST R. RANUCCI**

*State University of New York at Albany
Albany, New York*

GIVE Lewis Carroll credit for the first part of the title. The barber pole is my own idea. It's not the barber pole that intrigues me as much as the spiral wrapping on the pole. It's in the form of a *helix*. The easiest way to construct a helix is to begin with a rectangular sheet of paper—lined paper is quite appropriate (fig. 1a). $\overline{AC}$ is a diagonal of rectangle $ABCD$. To form the helix it will first be necessary to wrap the rectangle in the form of a right circular cylinder. Roll the paper so that $\overline{CD}$ can be pasted adjacent to $\overline{AB}$. A flap has been drawn in for this purpose. What was originally diagonal $\overline{AC}$ will then assume the form of a helix. It intersects each of the vertical segments (elements) at a constant angle. Point C will lie directly above point A. These points will be the extremities of an element. The helix constructed in this manner will make one complete circuit of the cylinder (fig. 1b).

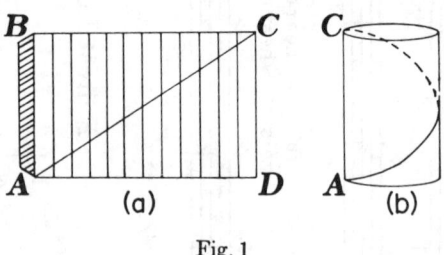

Fig. 1

If the helix is to make *two* circuits of the cylinder, the pattern shown in figure 2a can be used. This is simply a variation on the pattern shown before. Draw $\overline{EF}$ midway between $\overline{AD}$ and $\overline{BC}$. Construct *two* diagonals, $\overline{AF}$ and $\overline{EC}$. Form the cylinder as before and the helix will make two complete circuits (fig. 2b). In a similar

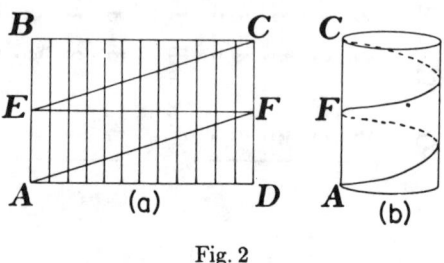

Fig. 2

manner, helixes can be drawn that will circle the basic cylinder three, four, five, or any number of times.

Bakst (1952) offers an interesting variation on the helix theme:

Suppose that you wish to construct a helix that will circle a cylinder *four* times. Lay out four adjacent, congruent rectangles on transparent plastic (fig. 3a). Draw segment $\overline{AC}$. Roll up the

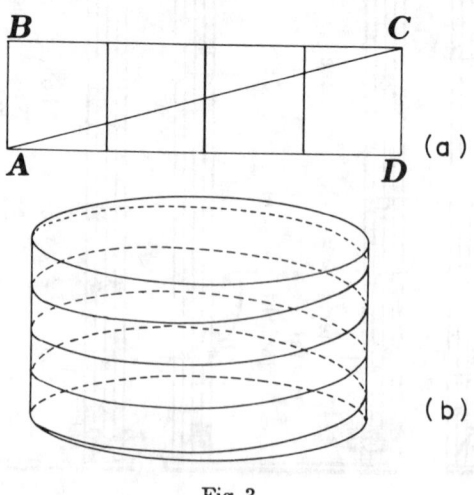

Fig. 3

entire assembly into one basic cylinder. This will mean, essentially, that *four* transparent cylinders will be superimposed. If C and A are made the extremities of an element, the four-fold helix can be viewed right through the transparent set of cylinders (fig. 3b). The process can be used for any number of circuits. As the number of rectangles increases, the inclination of $\overline{AC}$ will diminish. When the number of rectangles becomes sufficiently great, $\overline{AC}$ will tend to coincide with $\overline{AD}$; the branches of the helix will get closer and closer together (fig. 4).

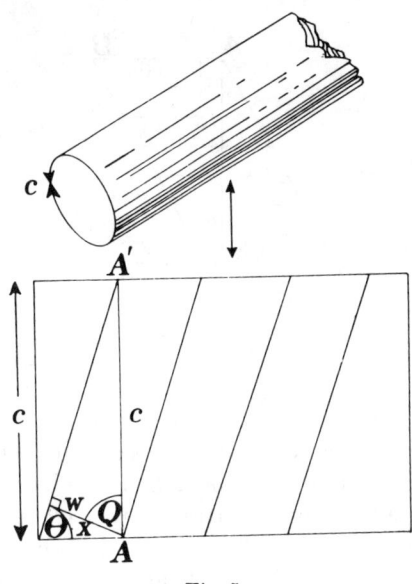

Fig. 5

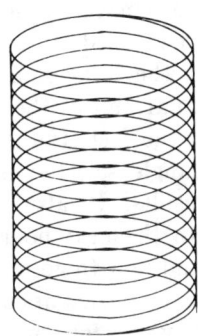

Fig. 4

From Barber Poles to Baseball Bats . . . .

Baseball players frequently wrap tape around bat handles for a more secure grip. Tennis players do the same thing with tennis rackets. The tape usually overlaps itself. Sometimes the overlap is pronounced; sometimes there is just a bare minimum. Under certain circumstances there would be no overlap, just tape showing, but no wood. When would such a thing occur? The rather peculiar shape of the bat handle makes mathematization of the problem difficult. It will be most convenient to deal with the handle in the form of a perfect circular cylinder. Figure 5 illustrates the situation encountered:

Let c represent the circumference of the right circular cylinder.

Let w represent the width of the tape being used.

Let θ represent the measure of the angle between an edge of the tape and an element of the cylinder.

$\overline{AA'}$ will be of length c since A and A' are, in reality, the same point. Angles θ and Q have the same measure since complements of the same angle, x, will be equal.

Thus: $\cos Q = \dfrac{w}{c}$ and $\theta = \cos^{-1} \dfrac{w}{c}$

If, for example, the circumference of the cylinder were to be 2 units and the width of the tape 1 unit, it would have to be laid down so that its edge is inclined 60° with an element of the cylinder: $\cos \theta = \tfrac{1}{2}$ and angle θ contains 60°. This means that no raw wood whatever will show. This means that no overlap whatever will occur. Few baseball players are aware of this, which may possibly account for the poor showing of some baseball teams.

From Baseball Bats to Paper Towel Rolls . . . .

The core around which paper towels are rolled is usually in the form of a right circular cylinder. It is most interestingly contrived. Such cores are based on parallelograms, such as ABCD, shown in fig. 6.

Mathematics in the World

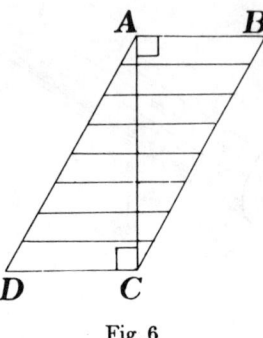

Fig. 6

Parallelogram $ABCD$ is unusual in at least one respect—diagonal $\overline{AC}$ is perpendicular to both $\overline{AB}$ and $\overline{CD}$. It is obvious that all segments parallel to $\overline{AB}$ and terminating in $\overline{AD}$ and $\overline{BC}$ will be of length AB (or CD). $\overline{AC}$ can be made the element of a cylinder. Both halves of the parallelogram can be rolled so as to form the cylinder itself. Its circumference will be of length AB. In commercial practice, two such parallelograms are used, one on top of the other but out of phase by 180°. This makes for greater strength. Figure 7 shows this:

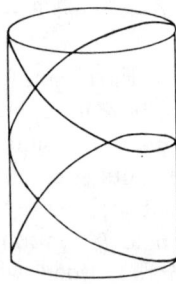

Fig. 7

The core of an average roll of paper towels will need more than one basic parallelogram for its formation (fig. 8). The

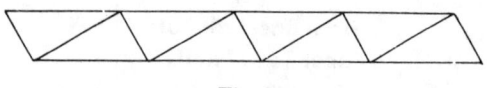

Fig. 8

procedure to be followed for such complex cores will be exactly as was previously described. Simply continue to roll the composite parallelograms round and round. In actual practice, mechanized equipment does this.

From Paper Towel Rolls to Circular Staircases....

Circular staircases offer an economical way of getting from one floor to another. The average set of stairs wastes a lot of usable floor-space. The circular staircase reduces this waste. The circular staircase is, of course, no panacea. Since the shape of the stair tread used in most of these staircases is in the form of a sector of a circle, the angular end of the tread offers little room for the feet. If the size of the basic central angle selected is increased, more room results. It also takes more turns to rise to the next floor level. Deciding where and how to go may well lead to vertigo. Efficient design of circular staircases calls, as usual, for intelligent compromises. The best way to get from one floor to the next is, in all likelihood, the ordinary ladder or the old-fashioned fireman's brass pole. You could slide gracefully from one floor to the one just below, but the lady of the house would probably not appreciate the advantages of either the ladder or the pole; neither would the man of the house if he had to transport a tea-tray loaded with Spode china.

The layout of the spiral staircase is fairly simple. To go from one floor to the next, a circular hole of convenient diameter is first cut out of the upper floor. The number of steps required to ascend the stairway depends on the distance between floors, the breadth of the tread, and the desired height of the steps. Most people find that a height of 8 inches is about right. If the distance between floors is eight feet (96″), twelve steps would be needed: $\frac{96}{8} = 12$. A convenient stair is formed when a central angle of 30° is selected. One complete circuit of the spiral would take a person from one floor to the next: $12 \times 30 = 360$. A sector-stair with a radius of 3 feet gives a chord-span of about 20 inches. This allows plenty of room for the average foot. It must be born in mind that *any* circular stair runs short of space the closer one gets to the central axis of the stair. Another of the problems faced in designing this type of stair has to do with head-room. In the stairway we have described, there is 8 feet between a tread and its recurrence directly above itself. Actually only 7 ft. 4 in. is usable, since the riser—the vertical part of the upper stair—gets in the way. We are speaking, of course, of stairways where more than one circuit is needed to get to an upper level.

Figure 9 illustrates two basic spiral

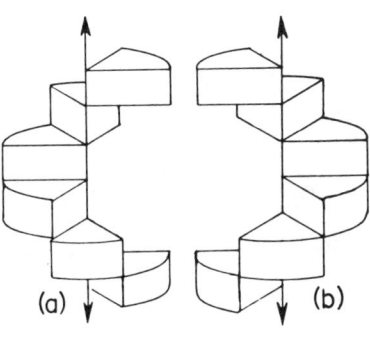

Fig. 9

staircases. One is right-handed and the other is left-handed. Both types can be found in old Scottish castles. There is a reason for them. Since a castle was built, ostensibly, for the protection of a resident noble, his own handedness would determine the style of spiral used. Figure 9a depicts one intended for the protection of a right-handed laird. If he were on the steps, fighting off the advances of an attacker, the noble would have relative freedom to wield his right-handed claymore. He would be standing on the wide part of the sector-step with freedom to operate. We shall assume that his attacker is also right-handed. The other staircase, shown in figure 9b, was intended for a left-handed noble. All bets would be off in the event that the attackers and the defenders had opposite handedness.

From Circular Staircases to Shoes—and Ships—and Sealing Wax

This article was deliberately written with tongue in cheek. I find that many a mathematics class is just too serious. My guess is that the pertinent anecdote, the amusing application of mathematics, the offbeat approach to the subject often pay off. Students sometimes remember vital things about our subject *because* they were approached in a lighthearted way. To this day I meet former students who have never forgotten that pencils, fundamentally, are either right-handed or left-handed. If the printing on an ordinary lead pencil reads from the lead-end toward the eraser, the pencil is right-handed. If the printing reads from the eraser down, the pencil is left-handed and intended to be read by a person holding the pencil in his left hand. Pencils become ambidextrous in two basic ways: either the printing on the pencil goes vertically, from the eraser down, or the pencil is sharpened at both ends. So it *isn't* an earth-shattering idea; the point is that the boys and girls remembered it....

REFERENCE

Bakst, Aaron. *Mathematics, Its Magic and Mystery.* New York: D. Van Nostrand, Co., 1952.

Part 3

And Then There Was Space!

Four articles were selected for this section. Spatial visualization and ways to help students visualize were part of almost every talk that Ernest Ranucci gave to teachers. In "The Weequahic Configuration," from the *Mathematics Teacher*, Ranucci introduced the reader to orthographic projection and its uses in a geometry classroom. The next reading, "Drawing for Math Teachers Who Can't Draw," from *Updating Mathematics,* is a natural next step. To be truthful, it was a lot easier to observe Ranucci draw and try to imitate him than to read about how to draw, but this article has been useful for a generation of mathematics teachers who need to know how to draw but who did not know Ranucci. Despite the availability of beautiful diagrams on purchased transparencies, there is still something particularly constructive about the evolution of a complicated diagram from simple line sketches on the chalkboard. Ranucci's instructions make such diagram sketching possible.

"Schlegel Diagrams," from the *Journal of Recreational Mathematics,* is an interesting introduction to unicursal routes. Ranucci also interspersed the history of mathematics and tidbits of mathematical art and literature in this reading. Even though the article appeared in a journal dedicated to recreational mathematics, Ranucci correctly identified the topic as one appealing to students in secondary school and college geometry classes. The next selection, "Spatial Aspects of the Venn Diagram," appeared in the *New York State Mathematics Teachers Journal.* It is interesting that the "symbolic logic" aspect of the topic is presently one of the major strands of the new New York State high school curriculum, but at the time this article was published, Ranucci was using the "new math" topic of sets and showing the possible variation in the kinds of shapes used to represent combinations of sets. The final selection, "Topology—through the Alphabet," from the *Mathematics Teacher,* extends the topological concepts implicitly introduced in these last two articles. In a succinct and carefully crafted article, Ranucci introduced ideas suitable for students at elementary, middle, and high school levels. His valedictory, a topological signature and a topological prayer, is another tongue-in-cheek reminder to all teachers not to take themselves too seriously.

The Weequahic configuration

E. R. RANUCCI, *American High School, San Salvador, El Salvador.*
An exercise in visualization in three-space.

BACK IN the good old days before "solid geometry" was criticized as much as it is today, we used to do a lot with orthographic projection in the solid geometry classes. We did not do the precise drawings that were turned out in the mechanical drawing classes—that was not our intention. The purpose of the unit was to exercise spatial imagery and to give visual intuitions a workout.

The customary approach to orthographic projection came first. Here a three dimensional object is imagined to be enclosed within a transparent rectangular solid. *Figure* 1 shows the views that are called the front view, top view, and end view. These two dimensional views are always placed as shown in *Figure* 1. For *Figure* 1 the following are true:

$$AB = CD = EF = GH$$
$$AC = BD = JK$$
$$EG = FH = IJ$$

Many such exercises were given. In some cases the orthographic projections were given, a sketch of the object being required. Sometimes the object was drawn and the orthographic projections were required. Due consideration was given the dotted line as contrasted to the solid line (*Fig.* 2).

After many such exercises students graduated to more sophisticated types. In these, two views were given, the third being required. Two of our favorites follow (*Fig.* 3).

The problem that developed into the "Weequahic" configuration (this all hap-

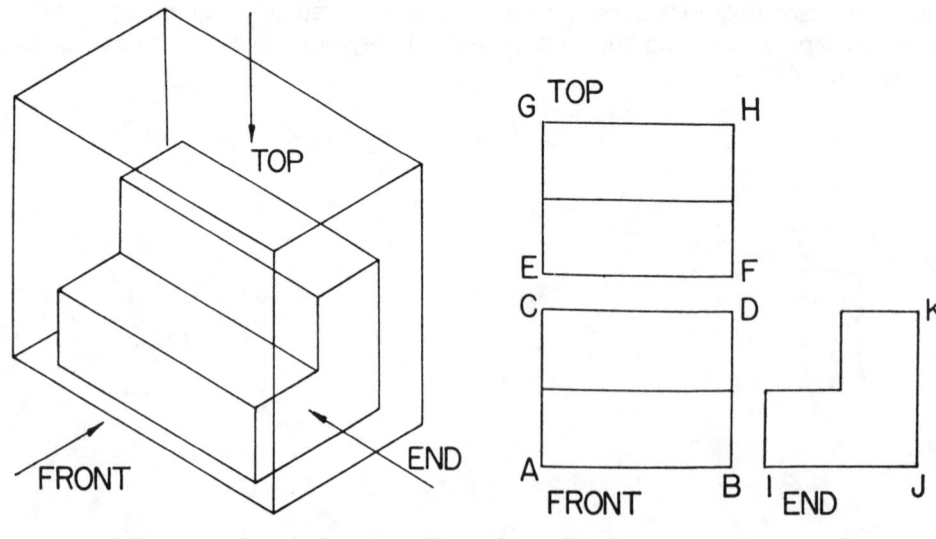

Figure 1

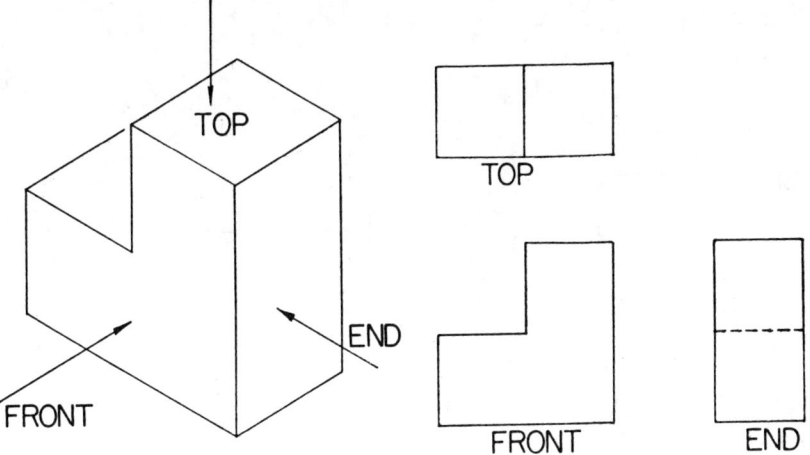

Figure 2

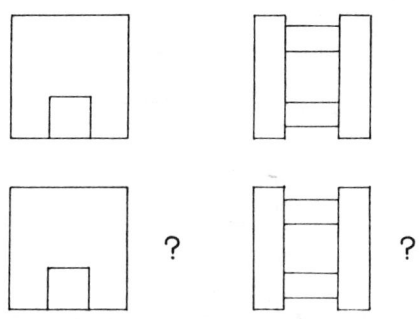

Figure 3

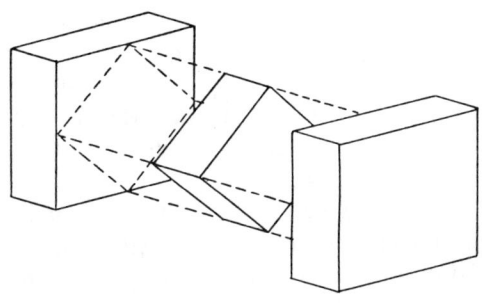

Figure 5

pened at Weequahic High School in Newark, New Jersey, between the years 1936 and 1956) started like this (*Fig. 4*):

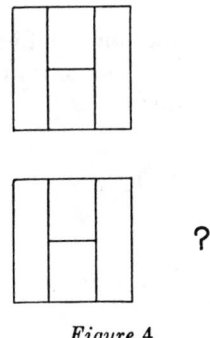

Figure 4

I had one object in mind (*Fig. 5*).

Little by little, however, students kept bringing in new solutions to the problem. At the last count, several *thousand* such configurations were shown to exist.

If the central figure in this "sandwich" is considered, here are variations on the possible form of the "filling" (*Fig. 6*).

Figure 6

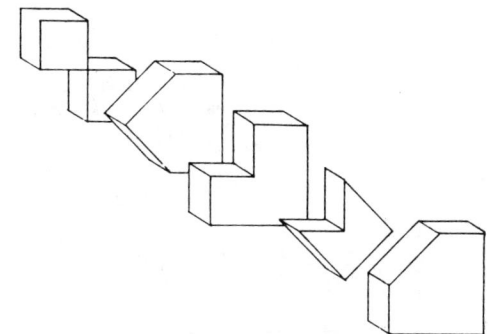

And Then There Was Space!

Figure 7

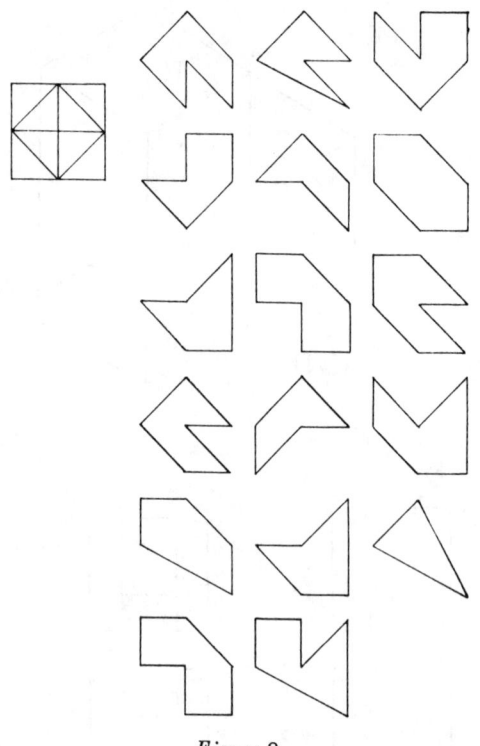

Figure 8

If the "bread" in the sandwich is considered, above are possible variations (*Fig. 7*).

The end view of the "filling" may be based on the following possible subdivisions of the square (*Fig. 8*). Any of the following outlines would show the requisite cross bars on the "H" needed in the center of the front and top views:

If only ten of the possible ends and ten of the central sections were placed in juxtaposition, 1000 such solids would be possible. Judicious interchange would increase this number considerably.

Some of the possible formations would be outlawed in cases where three adjoining faces were coplanar. This would be necessitated by the fact that when faces are coplanar no joining edges may be shown in orthographic projection.

The moral of the story is: have more than one possible drawing in mind in exercises of this type. If you do not, the students will probably show *you* that they exist.

If you would like more information on orthographic projection consult a text in mechanical drawing.

"The true mathematician is always a good deal of an artist, an architect, yes, of a poet."—
A. Pringsheim.

Drawing for Math Teachers Who Can't Draw

by Ernest. R. Ranucci

For teachers in the pitiable position stated in the title, there is but one basic rule: draw, draw, draw.

Half the battle consists in being willing to make the attempt, despite the poorly concealed snickers of an unappreciative class. Drawing, as dealt with here, is of the freehand variety. The niceties of mechanical drawing do not concern us. We are simply exploring the possibilities of the half-minute sketch which is as quickly erased. We seek to communicate an *idea* where, very often, the visual approach illuminates an abstraction not easily appreciated by the student. Not all of your charges understand the Pythagorean relation when it is expressed as the algebraic equation $c^2 = a^2 + b^2$. A large number of your group *may* get the idea through a sketch such as *figure 1*.

In this article we shall concern ourselves almost completely with blackboard drawing. The same principles, of course, apply to sketching in any medium.

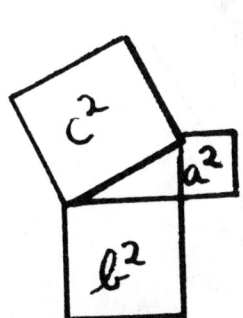

(Fig. 1)

RECTANGLES, SQUARES AND CUBES

The fundamental forms usually encountered in mathematics include the rectangle and the square. The rectangle is usually drawn as in *figure 2*. If a third dimensional viewpoint is

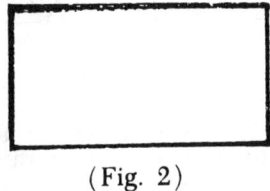

(Fig. 2)

desired, the rectangle is usually drawn as in *figure 3*. Orientation is facilitated when the front edges are drawn with the side of the chalk. The best point of view is usually presented when the basic angles formed with the horizontal are 30°—a characteristic of isometric drawing.

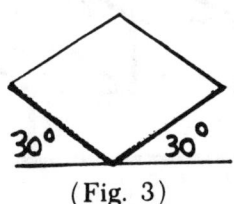

(Fig. 3)

Horizontal and vertical directions are extremely important in sketching. The blackboard ledge and divisions between adjacent boards are the chief visual cues in sketching geometric solids. They are responsible, in the most part, for determining whether or not a figure seems to be upright. The cube, for example, is frequently oriented so that the edges are parallel to a vertical direction. The angles which AB (*figure 4*) and BC make with a horizontal line

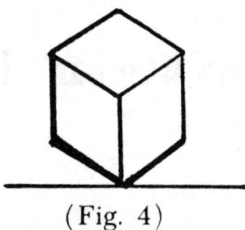

(Fig. 4)

determine the degree of tilt of the cube. By varying this angle, different views may be presented. (*See figure 5.*) The isometric drawing, *figure 5c*, is the view acceptable to most people.

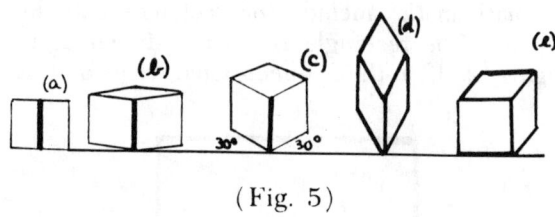

(Fig. 5)

Because of optic considerations, parallel lines often appear to meet. The rails of railroad tracks, for instance, appear to cross somewhere in the distance. In oblique drawing, however, parallel lines remain parallel regardless of the viewpoint of the observer. The front view of an object, in oblique drawing, is a true replica of the physical form itself. (*See figure 6.*)

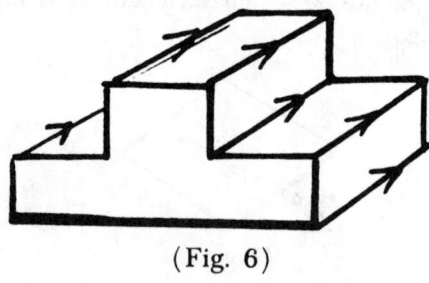

(Fig. 6)

In perspective drawing, on the other hand, parallel lines are regarded as meeting at "vanishing points." These points are determined by proximity to the object being represented, viewpoint of the observer and orientation of the sketch.

The cube, in one-point, or parallel perspective, is as shown in *figure 7*. Here vertical lines

remain vertical, horizontal lines remain horizontal, and the set of receding parallel lines passes through the vanishing point P.

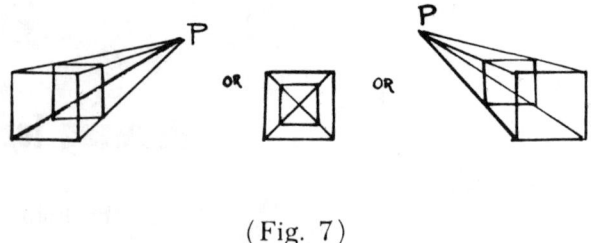

(Fig. 7)

In two-point perspective, the cube appears as in *figure 8*. This is the view we normally see. Here vertical lines remain vertical while

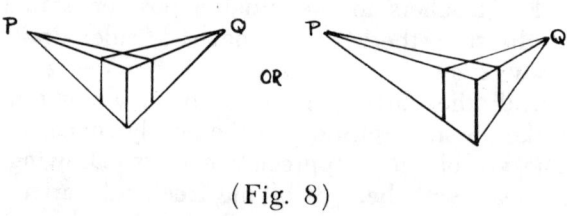

(Fig. 8)

each set of horizontal lines passes through a vanishing point (P and Q).

In three-point perspective, the cube takes on the appearance of *figure 9*. This is the view one might get of a skyscraper from quite a height.

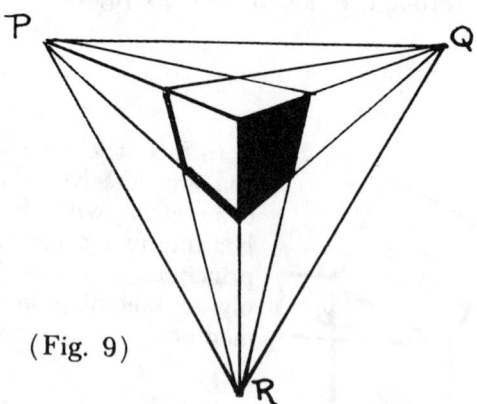

(Fig. 9)

The rectangular solid plays an important role in mathematics. Knowing how to draw it properly is worth-while. For most quick sketches, perspective is relatively unimportant. Some sketches, all based on the rectangular solid, are

Imaginative Ideas: Ranucci's Reservoir

shown in *figure 10*. Try drawing these and as many others as you can imagine.

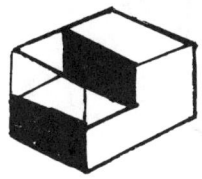  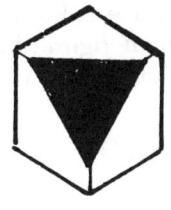 

(Fig. 10)

DIHEDRAL ANGLES

The dihedral angle (*see figure 11*) can be drawn easily by following these steps. If parallelism of lines is observed, the resulting diagram will appear "real." Notice that in the final drawing of the dihedral angle (*figure 11c*) lines AB, CD and EF are equal. This apparently

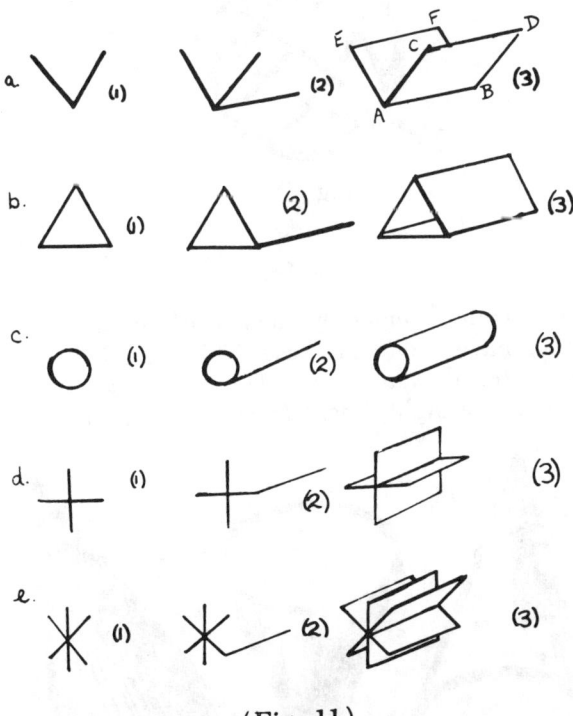

(Fig. 11)

minor matter results in a diagram which looks convincingly like the "actual" object.

The principle illustrated above and in the following diagrams is quite useful. Draw first the two-dimensional counterpart of the object to be represented. Then add one of the basic "depth" lines. Remaining "depth" lines should be parallel to the original one, and just as long.

Composite drawings which utilize these principles result in diagrams as shown in *figure 12*. Here, as in all the suggestions offered, many, many sketches are required before proficiency is realized. Try out your own ideas, using various media. The ball-point pen is especially useful for indicating shading.

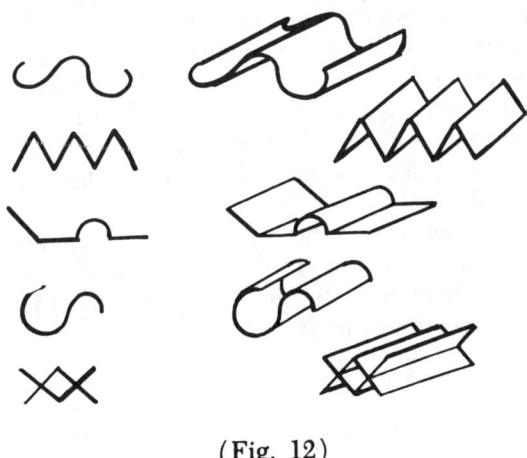

(Fig. 12)

CIRCLES AND SPHERES

The circle and its three-dimensional counterpart, the sphere, play an important role in mathematics. The blackboard sketch of the circle causes more difficulty than any other of the diagrams usually encountered in the teaching of mathematics. The secret of drawing a freehand circle reasonably well lies in the

And Then There Was Space!

freedom of the stroke. In drawing the circle at the blackboard, try the technique of using the elbow as the center of the circle. Start with the chalk at about "7 o'clock," and sketch a sweeping circle in a clockwise direction. If you can start as far back as "5 o'clock," the circle will probably end more smoothly.

The circle may be sketched on a small scale by using the technique depicted in *figure 13*. Tuck the forefinger and middle finger well un-

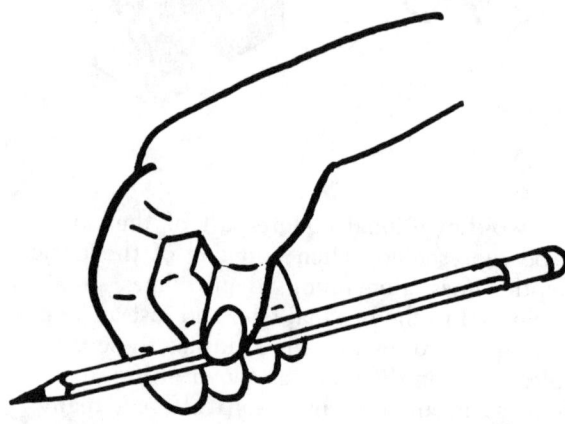

(Fig. 13)

derneath a sharp-pointed pencil, thumb on top. Keeping the point of the pencil in contact with the paper, turn the paper *underneath* the point of the pencil until the complete circle is traced out. In teaching this technique to children, it is well to have them take the paper out of their notebooks. It is rather unwieldy to turn a heavy, loose-leaf notebook.

The sketch of a sphere starts with a circle. Indications as to the third-dimensional nature of the sphere may be added by any of the devices illustrated in *figure 14*.

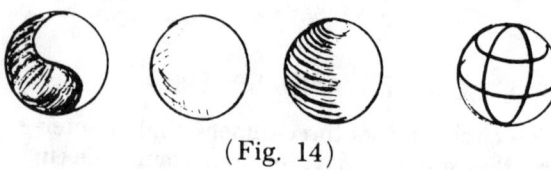

(Fig. 14)

The circle, viewed at an angle, is drawn as an oval or ellipse. This ellipse varies in width, according to the angle of vision, from the full diameter of the circle to zero. In every case, however, the length of the ellipse is that of the diameter of the original circle. (*See figure 15.*)

(Fig. 15)

Cross sections of a sphere are usually indicated as ellipses. (*See figure 16.*) These would include meridians, which govern longitude; and parallels, which govern latitude.

(Fig. 16)

CONES

The cones commonly dealt with in mathematics have a circular base. This base is usually sketched as an ellipse. The elements of the cone are straight lines. (*See figure 17.*)

(Fig. 17)

Imaginative Ideas: Ranucci's Reservoir

Schlegel Diagrams

Ernest R. Ranucci
State University of New York
Albany, New York

Schlegel diagrams are named for the 19th century German mathematician who invented them. A Schlegel diagram (you will also find it called *rabbatment*) is a two-dimensional diagrammatic device which seeks to preserve the essential nature of a three-dimensional structure. The important clerical features of a cube, for example, entail: 8 vertices, 12 edges, and 6 planes. Any of the diagrams in Figure 1 communicate these important statistics of the cube.

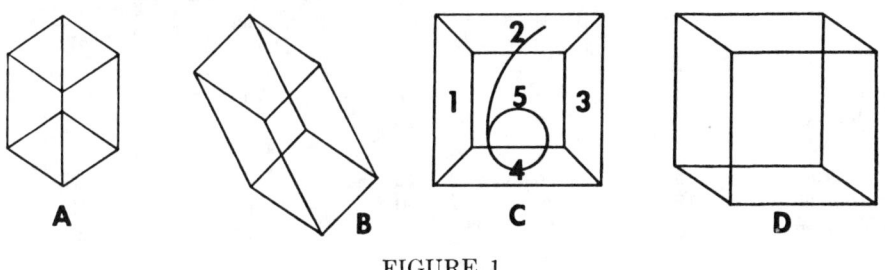

FIGURE 1.

Diagram C is superior to the others in one respect: no "accidental" vertices occur. Although distortion of physical features of the cube is bound to occur, the vital statistics of the cube are preserved. In this diagram—a Schlegel diagram—we are looking through the center of face 6 and observing the network formed.

Schlegel diagrams, and rabbatment in general, are used in branches of mathematics which involve relations among points, segments, and as models of physical situations. The fields of information theory, graph theory, and linear programming come to mind as studies in mathematics which utilize Schelegel diagrams.

Stein [4] refers to the *salesman* problem. Here we seek some route among n cities such that each is visited once and only once by a salesman. The problem as stated by Stein is analogous to the *Hamiltonian route*. This will be discussed later. Later on he refers to the *inspector* problem. Here segments in a network are being used to represent highways. The inspector wishes to scrutinize every section of his territory without examining a particular section twice. This problem is analogous to Euler's Unicursal Route. All of these problems make use of variants of the Schlegel diagram.

In Figure 2 the 8 vertices of a cube are used to represent 8 cities. *ABCDEFGH* is a representation of a salesman's route. It is one way of visiting all the towns in some order. There is no single inspector's route. It would take four separate trips to cover each of the twelve highways once and only once. This is indicated in Figure 3. The inspection cannot be accomplished in one traverse.

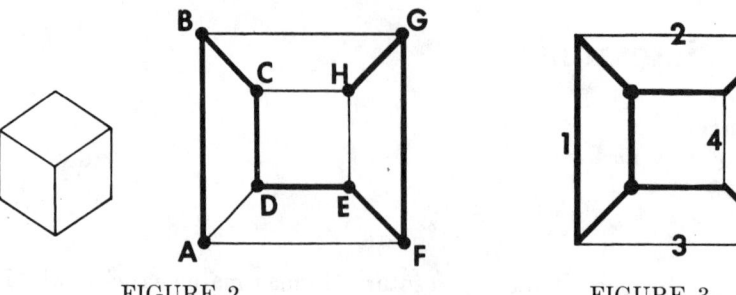

FIGURE 2.   FIGURE 3.

Rabbatment of the five regular solids provides a simple introduction to this field. The vital statistics pertinent to the five regular solids are given in Table 1.

TABLE 1

Solid	Faces	Vertices	Edges	Figure with Schlegel Diagram	Face Through Which Viewed
Tetrahedron	4	4	6	4	4
Hexahedron (Cube)	6	8	12	2	6
Octahedron	8	6	12	5	8
Dodecahedron	12	20	30	6	12
Icosahedron	20	12	30	7	20

Inspection of these statistics, preferably with models of the regular solids on hand, will disclose that the rabbatment lack in no detail. There is distortion, of course, but this is to be expected in a Schlegel diagram. After a certain amount of experimentation in the field of rabbatment you find yourself in a Picasso-Dali world, with overtones of *Alice in Wonderland* and a Gilbert and Sullivan fantasy (Things Are Not Just What They Seem). You will find the material extremely appealing to students in a class in geometry. It is specially suited to those students who react to the spatial world and who find the world of the dirty-subscript algebraist bereft of tangible appeal.

The rabbatment of the regular hexahedron (cube) has already been discussed. For each of the five regular solids a Hamiltonian route (salesman) is indicated in heavy lines. There are many others. In all cases each vertex is being visited once and once only. It is evident that with v separate vertices, $v - 1$ segments will have to be traversed. Mention was made originally of the value of the Schlegel diagram in determining a Hamiltonian route. The Hamiltonian Game, from which the field was derived, was invented by the Irish mathematician William Rowan Hamilton (1805–1865) in the 1850s. It consisted of a solid model of a regular dodecahedron or icosahedron. Brads were fastened at each of the vertices of these solids. The object of the game was to visit each of the "cities" represented by the vertices of the solid

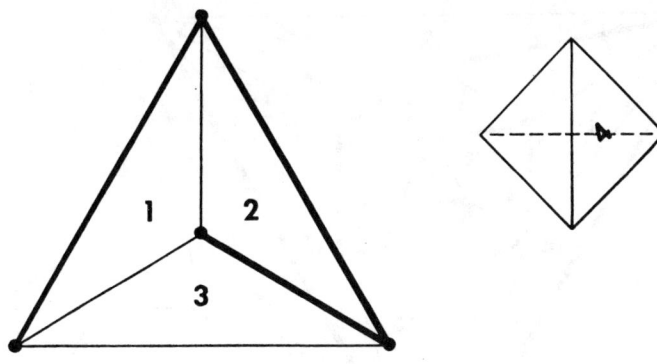

FIGURE 4.

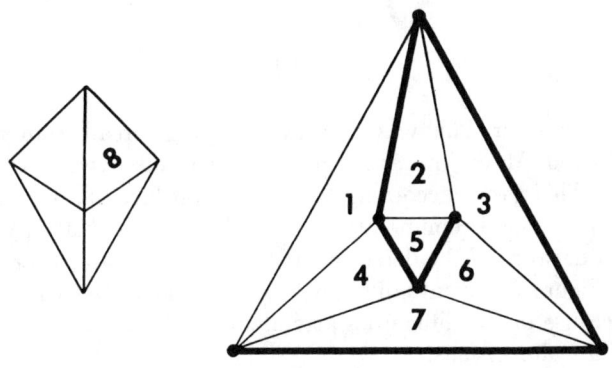

FIGURE 5.

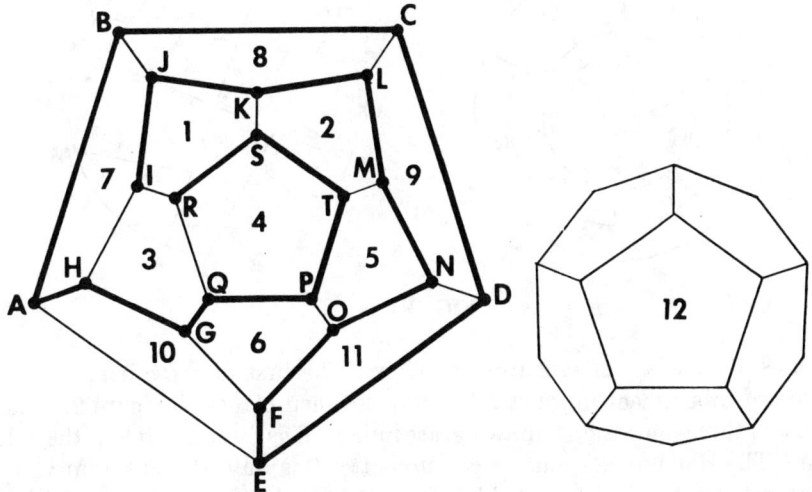

FIGURE 6.

And Then There Was Space! 51

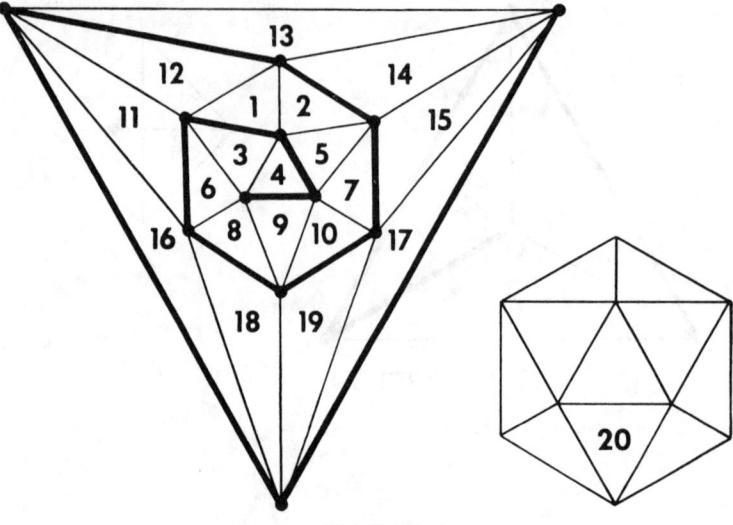

FIGURE 7.

in some particular order. This was done by wrapping thread from vertex to vertex in the order desired. Many interesting mathematical discoveries were possible with this recreation. This game, according to Martin Gardner [2], was the only one for which Hamilton received remuneration (£25—before devaluation). Ball [1] shows that the planar network of the dodecahedron is equivalent to either of the networks illustrated in Figure 8. Eventually the two-dimensional network superseded the solid model because of the difficulties attendant on the use of thread in describing a network in three dimensions.

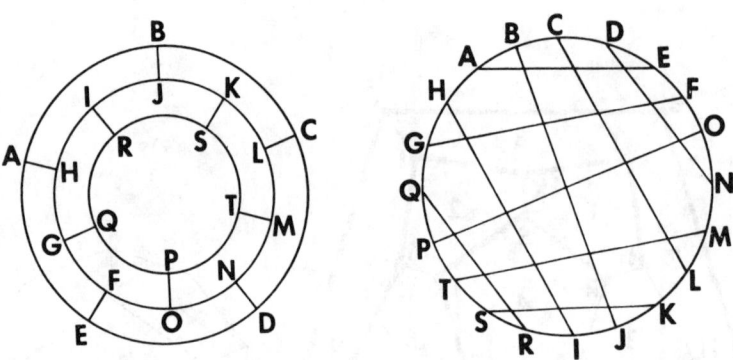

FIGURE 8.

A series of exercises in rabbatment follows. The first of these explains a fairly complicated rabbatment in some detail. The inclined planes in Figure 9, shaded in the original three-dimensional drawing, are indicated as faces 1 and 2 in the Schlegel diagram. The Hamiltonian route shows up better this way. If you squint your eyes in viewing the rabbatment, and think beautiful thoughts, you'll get a fairly clear picture of the three-dimensional values of the two-dimensional rabbatment. Lettering the vertices of the original drawing will also help. Every vertex in the rabbatment must repeat the essential structure of that vertex in the original.

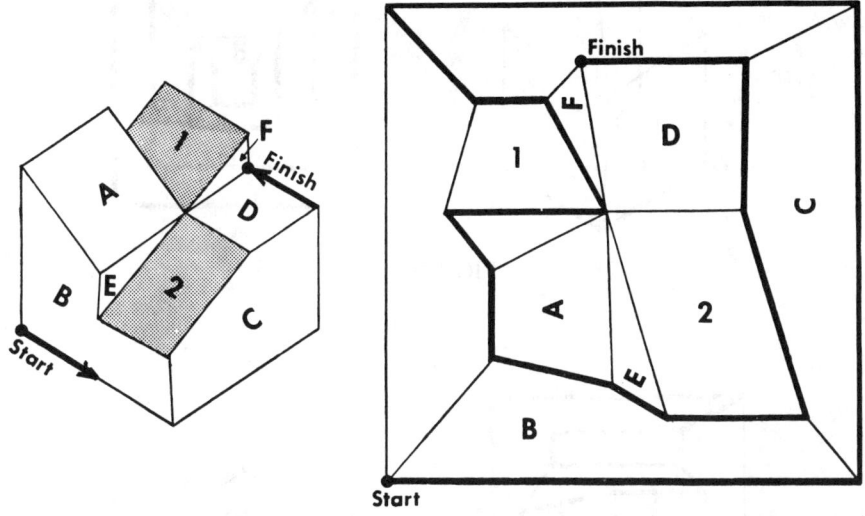

FIGURE 9.

The remaining exercises are explained only briefly (Figures 10 through 16). Later on in Figure 17 you will find a series of sketches where the rabbatment has not been provided.

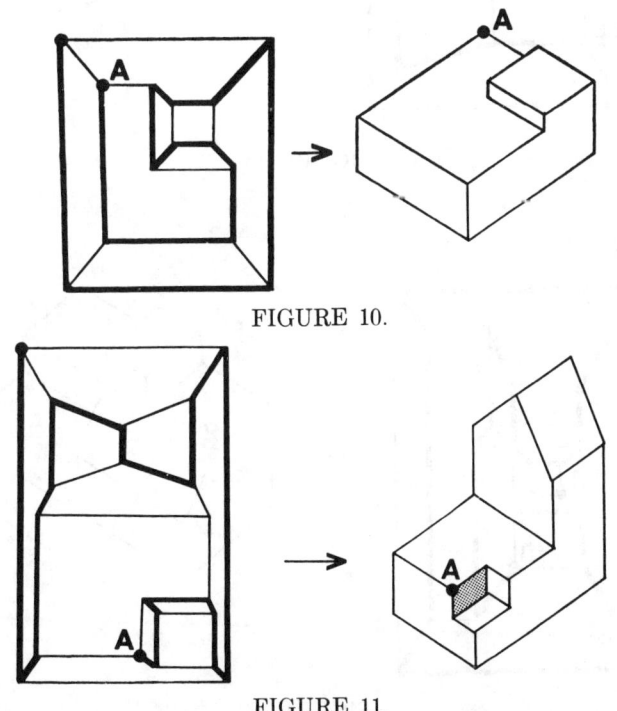

FIGURE 10.

FIGURE 11.

And Then There Was Space!

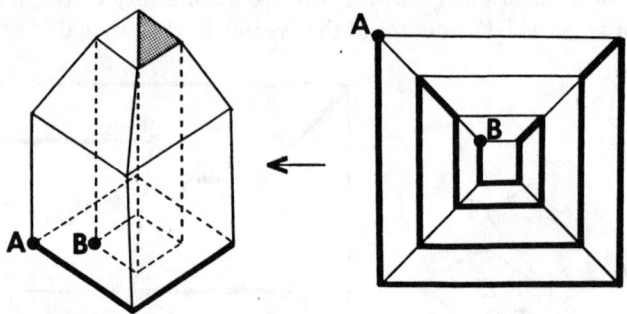

FIGURE 12.

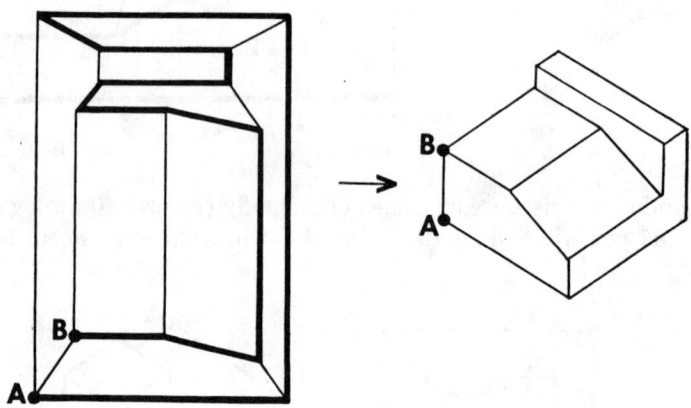

FIGURE 13.

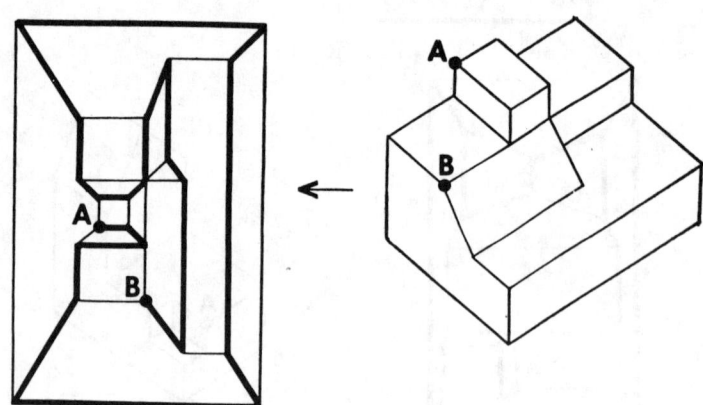

FIGURE 14.

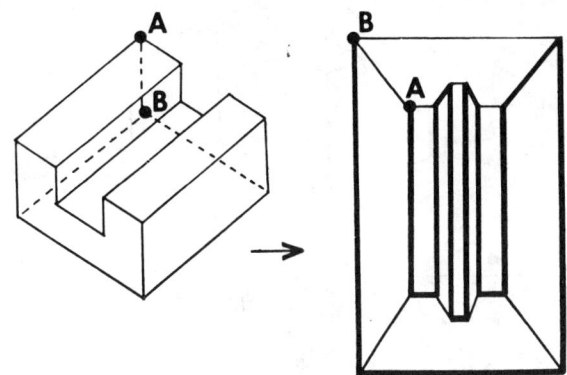

FIGURE 15.

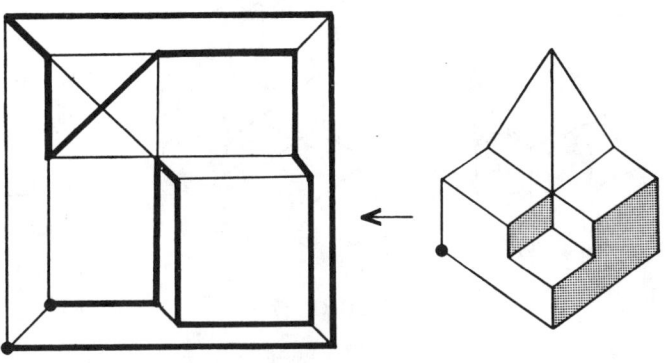

FIGURE 16.

All the Schlegel diagrams discussed pertain to a simple polyhedral surface. On such surfaces exactly two polygons meet at an edge (dihedral angle) [3]. It is always possible to go from one polygon to another by tracing some route along the edges of the original polyhedron. Such a polyhedron as that shown in Figure 18 is outlawed. This solid has holes in it that go all the way through. The Schlegel diagram would have to deal with two separate solids since there is no route connecting the edges of the two solids.

In actuality only the exterior form is solid. The interior form is a hole—a sort of three-dimensional cross. Come to think of it, Salvador Dali used it in one of his *Crucifixion* paintings. The pierced solid provides a stimulating exercise for calculation. If the cube is 3 x 3 x 3 and the tunnels are 1 x 1, located at the centers of the original faces, what is the volume, total surface, and perimeter of the pierced cube?

References

1. Ball, W. W. R., *Mathematical Recreations and Essays*, Macmillan, New York, 1960.
2. Gardner, M., *The Scientific American Book of Mathematical Puzzles and Diversions*, Simon & Schuster, New York, 1959.
3. Hilbert, D., & S. Cohn-Vossen, *Geometry and the Imagination*, Chelsea Publishing, 1956.
4. Stein, S. K., *Mathematics: The Man-Made Universe*, Freeman, San Francisco, 1963.

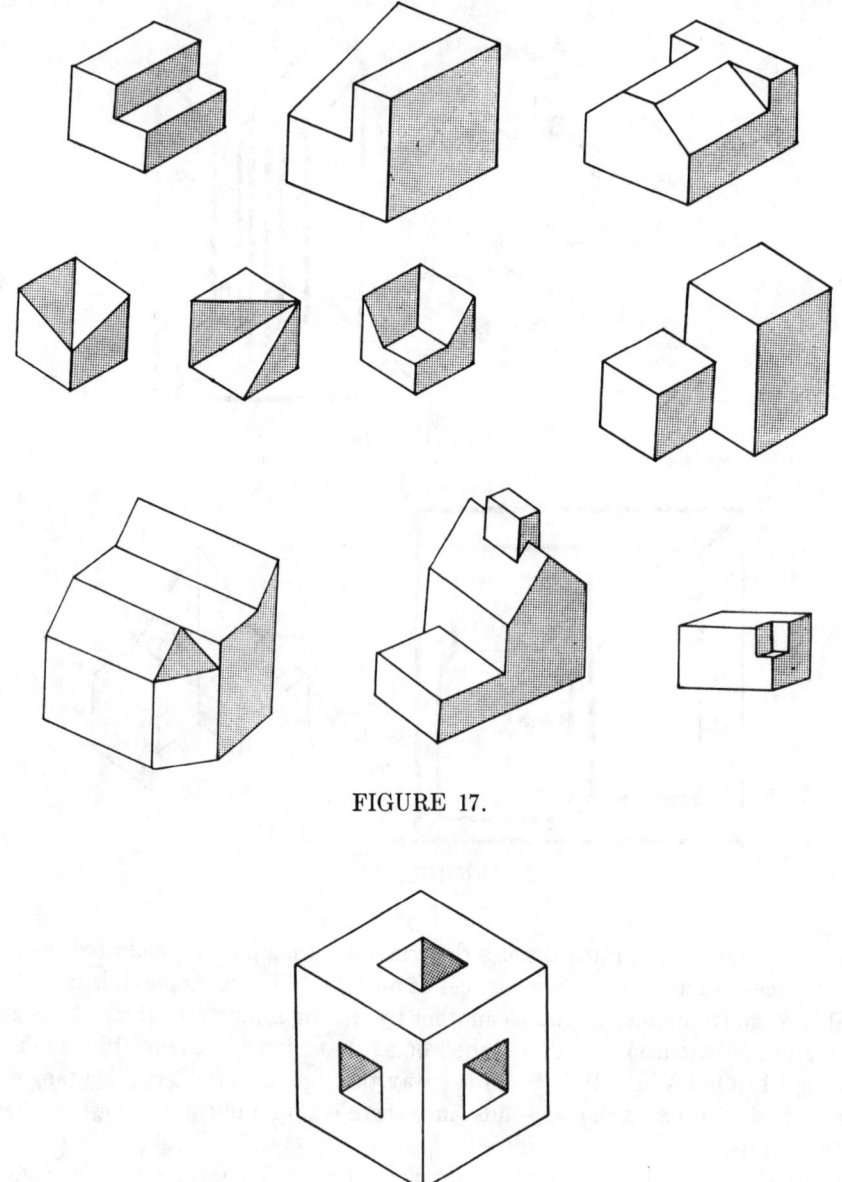

FIGURE 17.

FIGURE 18.

SPATIAL ASPECTS OF THE VENN DIAGRAM

Ernest R. Ranucci

State University of New York

Albany

The most useful device for organization of the complexities of set structure is the Venn diagram. The Venn diagram — a plane figure in which sets can be represented geometrically by letting points of the plane represent the members of the universe (U) — usually uses the circle for its symbolic representations.

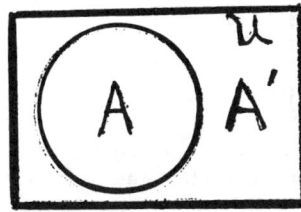

Thus: circle A contains all those elements present in A, and A', the complement of A, contains those elements within the universe U but not contained in A. Since Euclidean features of the circle have no bearing on the situation, any simple, closed, curve might have been used instead of the circle:

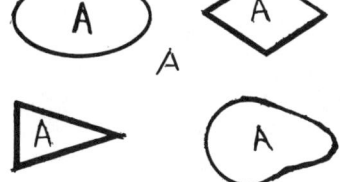

In all cases, A contains, symbolically, all elements of the set A.

When two sets are being considered, the Venn diagram usually involves two circles. These serve to divide the rectangle — symbolic of the collection of elements in the universe U — into four regions:

Region 1 contains those elements in A but not B

Region 2 contains those elements in B but not A

Region 3 contains those elements in both A and B

Region 4 contains those elements in neither A nor B

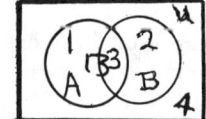

The use of the circle and rectangle have nothing to do with the basic set maneuvers. Any of the following diagrams might have been utilized:

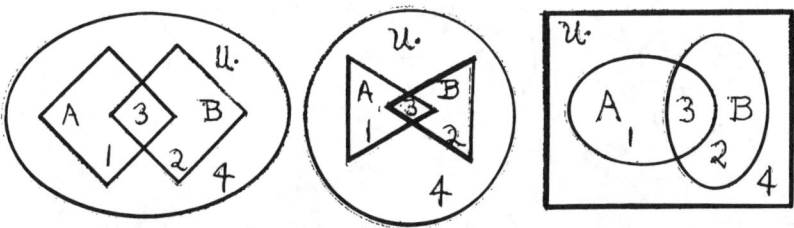

Reprinted with permission of *New York State Mathematics Teachers' Journal*.

And Then There Was Space!

If ~ is used to symbolize negation or denial, then the basic organization of each of the four regions "boils" down to:

1. A and ~B
2. B and ~A
3. A and B
4. ~A and ~B

Three set involvement usually results in a Venn diagram like that below:

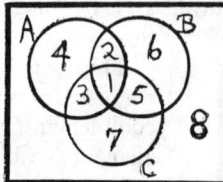

Here again, the use of the circle and rectangle is purely esthetic. Any other simple, closed, curve might have been used. Here the regions 1 to 8 represent sub-sets with the following considerations:*

1. A   B   C
2. A   B ~C
3. A ~B   C
4. A ~B ~C
5. ~A   B   C
6. ~A   B ~C
7. ~A ~B   C
8. ~A ~B ~C

Any of the Venn diagrams below might have been used to illustrate these relationships:

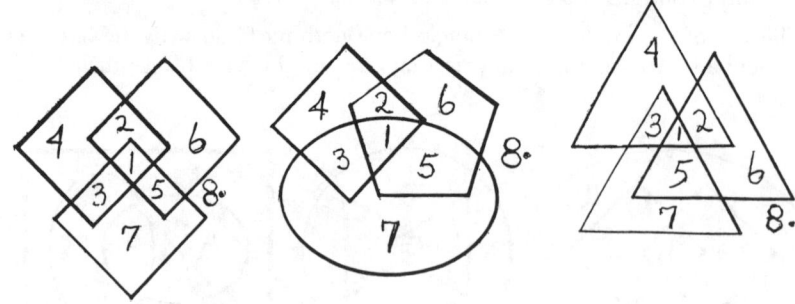

---

*For the sake of brevity, A and B and C shall be written as, A B C etc.

It is when more than three sets are being considered that circles lose their effectiveness. Four sets require provision for sixteen sub-sets: ($2^n = 2^4 = 16$). Four circles cannot partition two-dimensional space into sixteen regions. It is at this point that Euclidean considerations of the circle get in the way. The only feature of the circle that interests us in the Venn diagram is its simple, closed, curve property. The circle separates the plane on which it is drawn into two basic regions — an inside and an outside. *Any* simple, closed, curve will do this. Other features of the circle, with regard to its locus properties, have no value in the Venn diagram.

Four sets will, then, require use of configurations other than the circle. The sixteen sub-sets required in the case of the four sets are as follows:

1. A   B   C   D
2. A   B   C ~ D
3. A   B ~ C   D
4. A   B ~ C ~ D
5. A ~ B   C   D
6. A ~ B   C ~ D
7. A ~ B ~ C   D
8. A ~ B ~ C ~ D
9. ~ A   B   C   D
10. ~ A   B   C ~ D
11. ~ A   B ~ C   D
12. ~ A   B ~ C ~ D
13. ~ A ~ B   C   D
14. ~ A ~ B   C ~ D
15. ~ A ~ B ~ C   D
16. ~ A ~ B ~ C ~ D

Any of the following Venn diagrams will suffice:

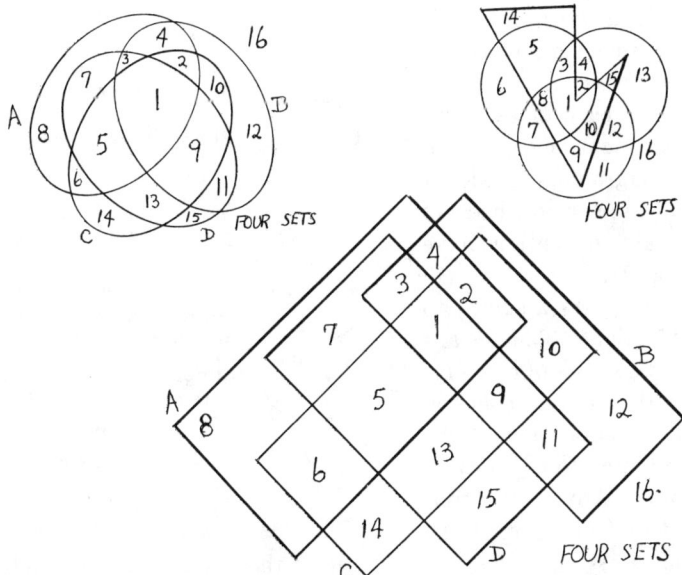

And Then There Was Space!

TOPOLOGY—through the Alphabet

By ERNEST R. RANUCCI

State University of New York
Albany, New York

VISUAL aspects of topology rate rather low in the lexicon of the topologists. In fact, visual aspects of *anything* rate rather low with the mathematician. It is, however, the very naïveté of a visual approach that fascinates the young student of mathematics. An imaginative teacher can, from time to time, enrich the teaching of geometry with topological gleanings.

This article concerns itself with the uses that can be made of the letters of the alphabet with regard to elementary topological considerations. A bibliography has been provided for those teachers who would like to investigate basic notions of topology.

The study of topological equivalence is basic to the study of topology. In essence, two or more configurations are topologically equivalent (homeomorphic) if the structures of each are identical. A one-to-one correspondence is always involved; the neighborhoods of each pair of corresponding points look identical.

The configurations in each of the rows shown in figure 1 are homeomorphic. In each case, a Venn diagram symbolic of the essential structures in each of the rows has been drawn. Certain conventions have been adopted (see fig. 2).

In no cases are the figures assumed to be embedded in the plane. This means that if a configuration consists of two simple closed curves with but one common point, its physical representation could be one of many. It is unimportant *which* of the possibilities is used. Any of the configurations in row B of figure 1 could be used to represent this class of curves.

The alphabet by its very familiarity is a suitable vehicle for the demonstration of homeomorphisms. The design of the letters used is extremely important. Serifs —the smaller lines used to finish off cer-

Fig. 1

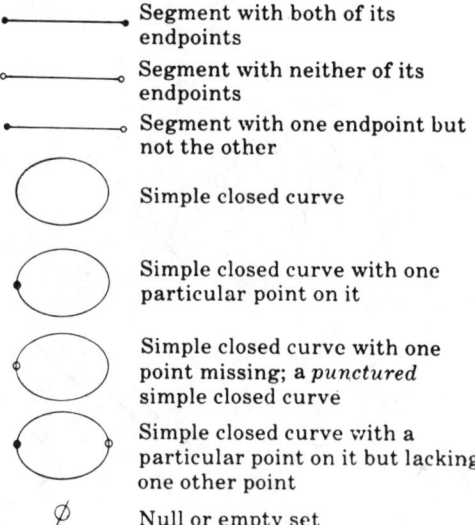

Fig. 2

tain of the letters—can completely change the essential topological nature of letters. In figure 3 the alphabet has been separated into topologically equivalent families. Serifs have *not* been used. In each case, a suitable symbol has been invented to indicate the structure of letters within the family. A Venn diagram has also been suggested.

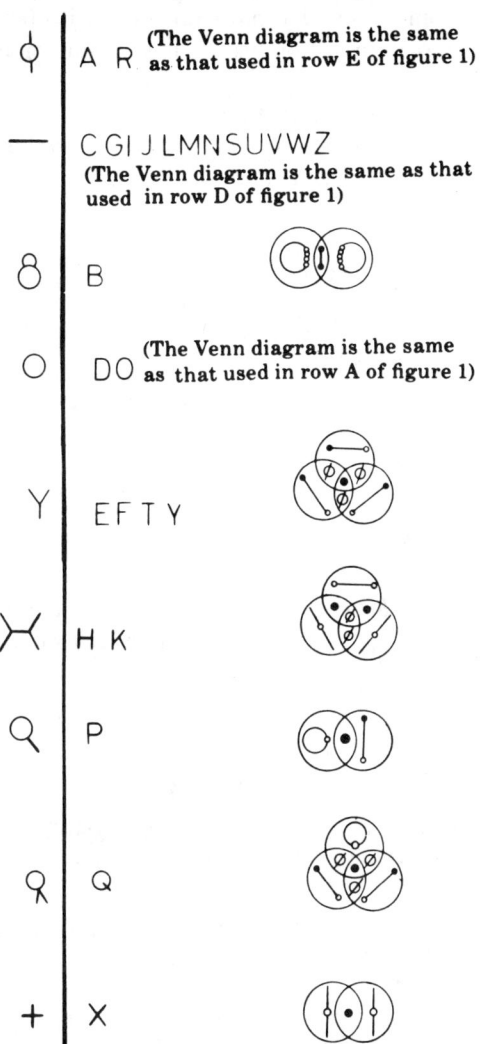

Fig. 3. In cases where *more* than three sets of points *could* have been involved, a simple Venn diagram of no more than three sets of points has been chosen. When more than three sets of points are encountered, circles can no longer be used in a Venn diagram because of spatial limitations. In such cases, simple closed curves other than circles can be employed in their stead.

Some further explanation might be helpful in considering the letters included within the Y family. (See fig. 4.) If some convenient point O is selected, it is evident that a one-to-one correspondence can be established between the points that constitute angle RST and those that make up arc PQ. Whether or not the arc is larger or smaller than the lengths of the sides that form the angle is unimportant. After all, metric considerations belong to Eu-

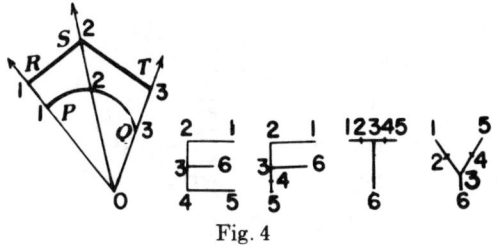

Fig. 4

clidean geometry, not to topology. A segment one-quarter of an inch long is, topologically speaking, homeomorphic to a segment one-quarter of a billion miles long. Each has an initial and a terminal point, and each contains exactly the same number of points within its extremities. Consequently, the letters E, F, T, and Y can be considered homeomorphic, as will be evident by an examination of figure 4: Point 2 is between points 1 and 3 in every case. The strategic point 3 is between points 2 and 4 in all cases.

It would be profitable to find out what other families would have to be added were the letters in the Greek alphabet added to our collection. See figure 5. The Russian (Cyrillic) alphabet can also be analyzed, as shown in figure 6. (Certain little-used letters have been eliminated.)

Entire words can be homeomorphic. Such words are made up of homeomorphic letters—for example, ART, RAY, RAT. Words can be homeomorphic even though the topologically equivalent letters do not occur in the same order within the words —for example, HOUSE, HIDES; CLEFT, TITLE; WAVE, SURF; and BUTTER, BREEZY. Sometimes a slight bit of poetic license can be applied. I choose to call this "topological tampering."

NEWS = ITEM

NEWS ITEM

MAPLE = SUGAR

MAPLE SUGAR

JUDGES = JURIES

JUDGES JURIES

It would be interesting to find out what topologically equivalent sentences could

And Then There Was Space!

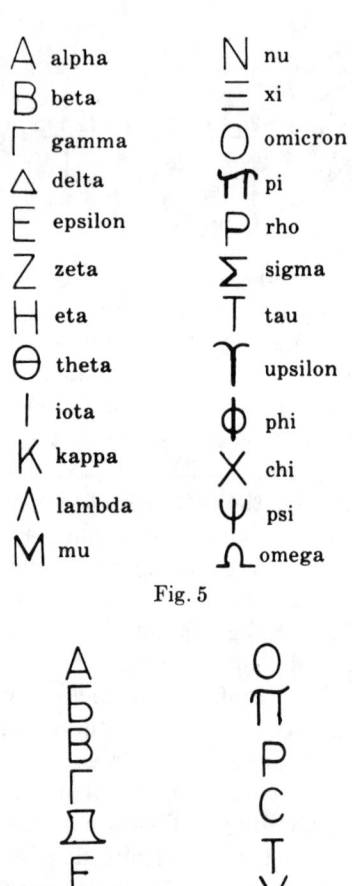

Fig. 5

Fig. 6

or senior high school levels. There are naïve ideas in elementary topology that students can easily comprehend. As the ideas of topology expand, the field becomes more and more abstract, with fewer possibilities for use at lower educational levels.

I would like to terminate this article by including a quotation (in symbols) that offers a sage bit of advice. It is called "Prayer of the educator":

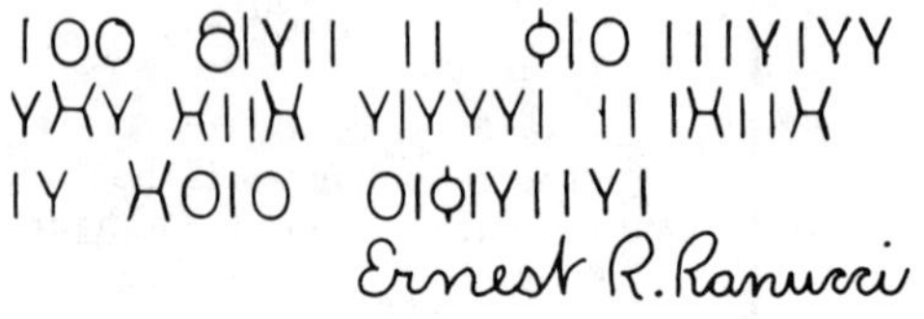

IOO 8IYII II ΦIO IIIYIYY
YXY XIIX YIYYYI II IXIIX
IY XOIO OIΦIYIIYI

Ernest R. Ranucci

His topological signature:

Translation: "God bless us and justify the high esteem in which we hold ourselves"

be invented. It would also stimulate class discussion to see who could find the largest pair of words that are homeomorphic. I myself would like to know.

This article has been deliberately kept at a level suited to the needs of the students in the upper elementary, junior high,

REFERENCES

Bergamini, David. "Topology: The Mathematics of Distortion." In *Mathematics*. Life Science Library series. New York: Time-Life Books, 1959.

Fletcher, T. J. *Some Lessons in Mathematics*. Chap. 8. Cambridge: At the University Press, 1965.

Kasner, Edward, and James Newman. *Mathematics and the Imagination*. Chap. 8. New York: Simon & Schuster, 1940.

May, Kenneth O. "The Origin of the Four-Color Conjecture." MATHEMATICS TEACHER 60 (May 1967): 516–19.

Ranucci, Ernest R. "Topology of the Alphabet." In *Updating Mathematics*, teacher's ed.

Tammadge, Alan. "Networks." MATHEMATICS TEACHER 59 (November 1966): 624–30.

Tucker, Albert W., and Herbert S. Bailey. "Topology." *Scientific American*, January 1950.

Wilder, R. L. "Topology: Its Nature and Significance." MATHEMATICS TEACHER 55 (October 1962): 462–75.

Part 4

Inventiveness in Geometry

Ranucci's favorite subject was geometry, and he particularly enjoyed creating new chains of definitions or investigating existing ones. The first two articles in this section are examples of his inventiveness at work. In "On Skewed Regular Polygons," from the *Mathematics Teacher,* Ranucci used analogy to help the reader compare the more familiar polygon to the various skewed polygons he illustrates. He used the second article, "The Congruency of Quadrilaterals," from the British journal *Mathematics Teaching,* to provide a reasonable definition of congruent quadrilaterals and through this specific example to illustrate the inventive nature of mathematics. The final two articles in this part deal with tessellations—"A Tiny Treasury of Tessellations" and "Master of Tessellations: M. C. Escher, 1898–1972." Both articles appeared in different issues of the *Mathematics Teacher.* In the first, Ranucci demonstrated the way any teacher or student could create tessellations—a skill he expounded on beautifully in a later book written with J. L. Teeter. The second article is a brief introduction to the life and works of Escher—an example of a human aspect of mathematics that might readily be used in the classroom.

A polygon is regular if its sides are equal and its interior angles are equal. But what if the polygon's elements are not coplanar?

ON SKEWED REGULAR POLYGONS

By ERNEST R. RANUCCI
State University of New York
Albany, New York

THE regular polygons considered in the definition above are those encountered in plane geometry. These are configurations whose elements are coplanar. Polygons whose elements are *not* coplanar are termed *skewed*. The triangle, by its very nature, must be a plane configuration (fig. 1). (Three noncollinear points always determine a plane.)

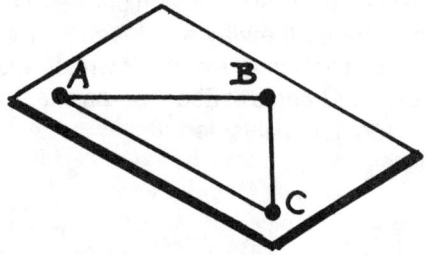

FIGURE 1

The quadrilateral, on the other hand, need not be planar (fig. 2). If its diagonals AC and BD intersect, it must be planar. (Two intersecting lines determine a plane.)

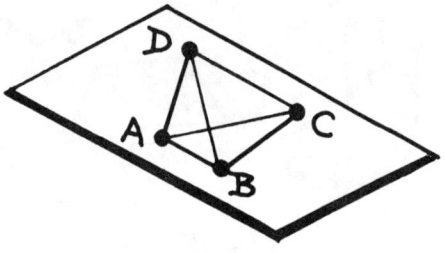

FIGURE 2

If there is no point common to the two diagonals, the quadrilateral must be skewed (fig. 3).

Regularity of polygons implies both equiangular and equilateral qualities. Thus, a *skewed* regular quadrilateral, analogous to the square in plane geometry,

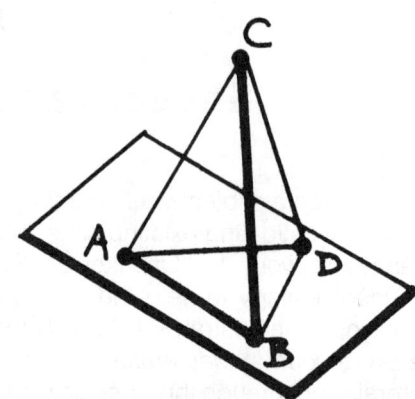

FIGURE 3

is formed by four of the six edges of a regular tetrahedron—a polyhedron all of whose four faces are congruent equilateral triangles. To form this quadrilateral, simply eliminate two of the skewed edges of the solid (fig. 4).

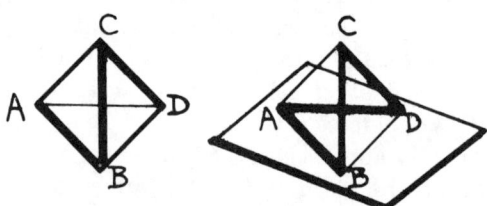

FIGURE 4

All squares are identical, except for scale. We term them "similar." There is but one basic regular quadrilateral in plane geometry. Peculiar to skewed regular polygons (hereafter referred to as SRP) is the fact that there may be more than one type within its classification. The plausibility of this statement may be checked by examining the following development (fig. 5).

Let plane 1 be parallel to plane 2. Circles O and O' are bases of a right circular cylinder. In each of the circles perpendicular diameters are drawn. The diameters of

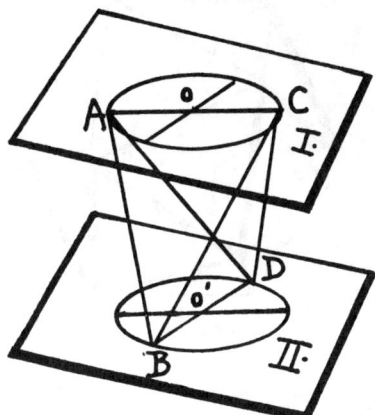

FIGURE 5

lower circle is "out of phase" with the one in the top circle. Its vertices have been advanced along 60° of arc. Alternate joining of vertices produces the desired SRP.

FIGURE 7

one circle are parallel to those of the other. *ABCD-A* is an SRP. Its vertex angles, none of which are coplanar, need not be 60° as in the previous case, which related to the surface of a regular tetrahedron. As the distance between the parallel planes increases (or as the radii of the circles are decreased) the angles of the SRP approach 0° and, ultimately, the "icicles" formed approach four superimposed line segments. At every stage of the proceedings the quadrilateral is regular. The sum of its four vertex angles varies, unlike the constant sum of the angles of a quadrilateral in plane geometry. As the planes approach each other, other factors being fixed, the SRP degenerates into . . . ?

Another approach to the skewed regular quadrilateral is through the rhombus (fig. 6). If *ABCD* is folded on its shorter diagonal, as indicated, the angles at *B* and *D* can be so altered as to equal those at *A* and *C*. As angle *A* is altered to approach 0°, so do the other three. Thus the result is a degenerate fourfold segment.

The cylinder approach to the generation of SRPs is richer in the possibilities it offers than that of the rhombus derivation. Consider the skewed regular hexagon in figure 7. Equilateral triangles were inscribed in both of the circles. That in the

It is evident that the cylinder generation produces only SRPs with an even number of edges. It is also evident that there are many variations of SRPs within their own classification and that there is no constancy whatever in the sum of the vertex angles of such SRPs.

There are many SRPs whose generation is not connected with that of the approach via a cylinder. Many of these are related to the twelve edges of a cube and variations on this theme. Consider the skewed regular polygons in figure 8.

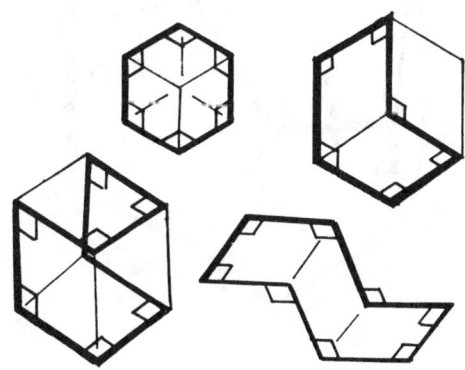

FIGURE 8

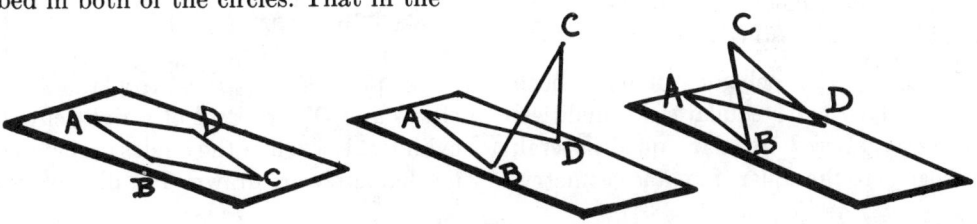

FIGURE 6

Inventiveness in Geometry

65

Other SRPs may be derived from selected edges of the regular octahedron (fig. 9). Still others come from the regular dodecahedron (fig. 10) and icosahedron (fig. 11). It is evident that there is a wealth of possibilities in these configurations of twelve and twenty faces respectively, when the number of edges of each, i.e., 30, is considered.

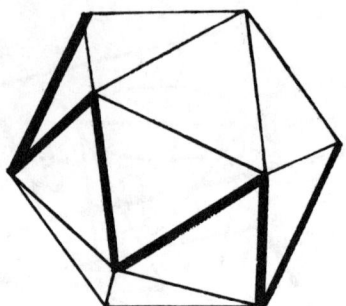

FIGURE 11

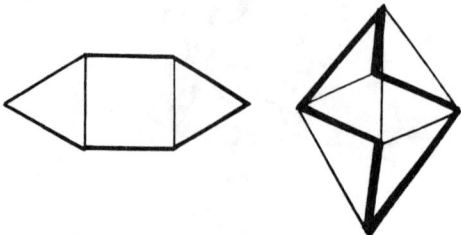

FIGURE 9

Figure 10 illustrates, for the first time in our discussion, an SRP with an odd number of edges. Polygon $ABCDEFGHIJK$–A has eleven equal edges. Each of its vertex angles is 108°.

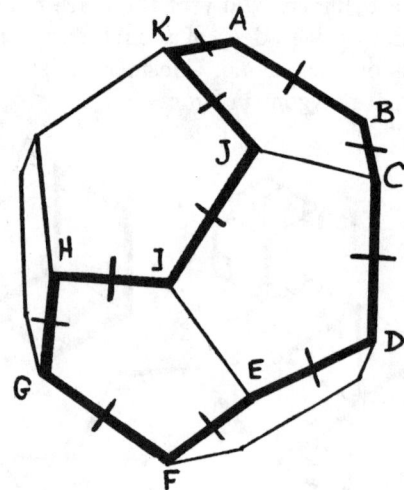

FIGURE 10

Figure 11 illustrates six of the ten edges of a regular decagon. Reference must be made to a solid model to find the other four edges. Interestingly enough, this decagon may also be generated by the cylinder approach.

The diagrams in Figure 12 illustrate the plausibility of the existence of an SRP of seven edges. If right triangle AGB were folded along hypotenuse GB, the angle AGF would change from 123° to 33°. Evidently the angle would have to pass through 90° to achieve this transition. In a similar manner the folding of trapezoid $FGBC$ along FC would result in the transition of angle GFE from one of 192° to one of 12°. Here again the size of angle GFE must be 90° at one point of the transition. Consequently the skewed regular heptagon $ABCDEFG$–A must exist. In this case its vertex angles are 90°.

The result of this bit of investigation into the field of skewed polygons is that the regular polygon is not the province of two-dimensional geometry only. Skewed regular polygons exist in three dimensions. The number of sides may be odd or even. There is more than one type of SRP within its own classification, unlike the situation that exists in plane geometry.

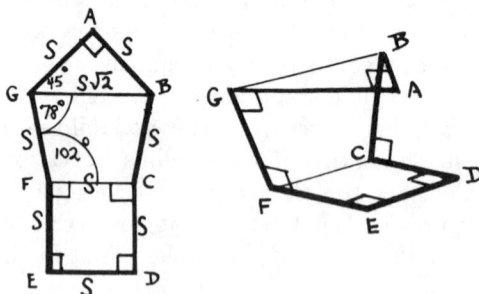

FIGURE 12

The Congruency of Quadrilaterals

ERNEST R. RANUCCI
State University of New York at Albany

Much has been written on the congruency of triangles. Textbooks in plane geometry couldn't do without it. Relatively little has been written on the congruency of quadrilaterals. A lesson or two on the subject usually illuminates congruency of triangles by pointing out likenesses and differences between the two. A discussion on the rigidity of polygons in general will serve as a good jumping off place.

The triangle is the only polygon which is rigid by itself. This means that segments AB, BC and AC, properly selected, will determine one and only one shape—that of the triangle itself. By '*properly selected*' we mean that the sum of the lengths of any two sides must exceed the third. This is equivalent to saying that the sum of the two shortest sides must exceed the longest side. If force were to be applied at A (or B or C), the other vertices being fixed, the only change which could come about would arise if one or more of the segments were to be deformed in some way.

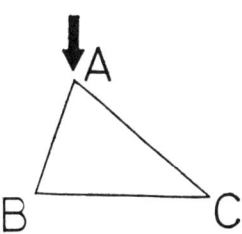

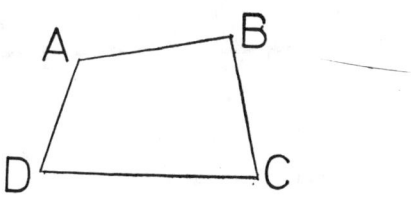

If a quadrilateral $ABCD$ were to be formed by four properly selected segments, deformity could occur in many ways. Quadrilaterals 1, 2, 3 and 4 all contain the same four segments as the original. No *unique* quadrilateral is formed under these conditions. (What constitutes *properly selected segments* in the case of the quadrilateral?)

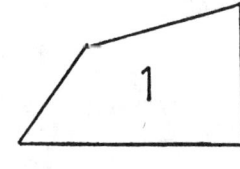
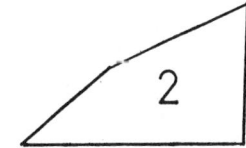
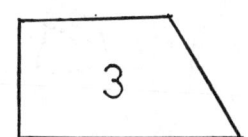
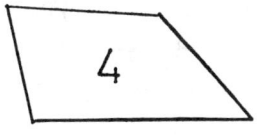

If one or more diagonals were to be added to the quadrilateral rigidity *would* occur. Now the quadrilateral is made up of triangles. Each is rigid. Greatest strength would occur if both diagonals were drawn, then fastened to each other at E. Now *eight* triangles strive for rigidity. (Identify the eight.)

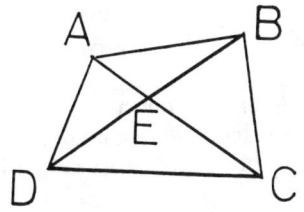

Inventiveness in Geometry

Congruency theorems pertinent to quadrilaterals almost invariably involve **considerations of rigidity.** This is why a preliminary study of this aspect of geometry is needed.

In the diagrams which follow, theorems on congruency of triangles are paralleled (where pertinent) by comparable theorems on congruency of quadrilaterals; only convex quadrilaterals are considered.

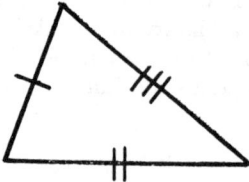 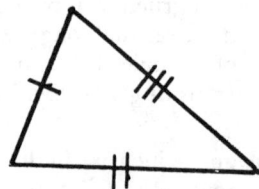

Two triangles are congruent if three sides of one equal three sides of the other (SSS).

No such theorem exists for quadrilaterals.

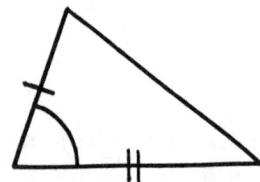 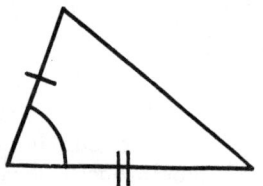 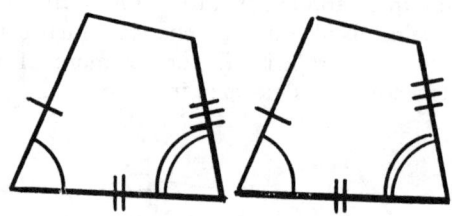

Two triangles are congruent if two sides and the included angle of one equal two sides and the included angle of the other (SAS).

Two quadrilaterals are congruent if three sides and two included angles of one respectively equal three sides and two included angles of the other (SASAS).

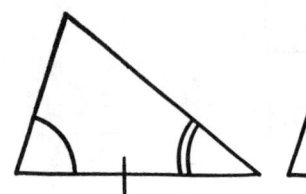 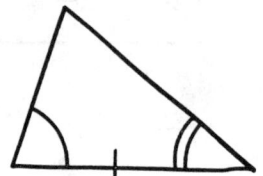  

Two triangles are congruent if two angles and the included side of one equal two angles and the included side of the other (ASA).

The angle-side-angle congruency is usually followed by the side-angle-angle congruency. After all, the sum of the angles of a triangle is constant.

Two quadrilaterals are congruent if three angles and two included sides of one respectively equal three angles and two included sides of the other (ASASA).

No such theorem holds for quadrilaterals. If three angles of one quadrilateral equal three of another, the fourth angles must be equal. But quadrilaterals are not necessarily congruent if one side and four angles of one equal respectively a side and four angles of another.

Imaginative Ideas: Ranucci's Reservoir

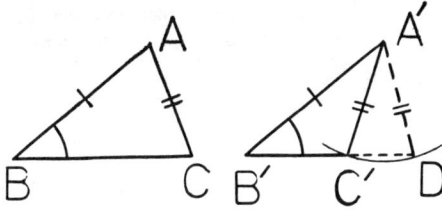 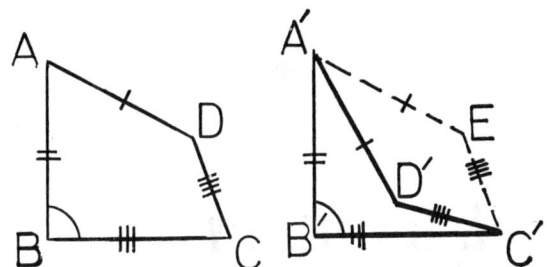

The ambiguous case normally arises if two sides and an angle opposite one of them are involved. Thus triangle *ABC* above is not congruent to triangle *A'B'C'*. If length *A'C'* were to be transferred to position *A'D*, congruency *would* result. Hence the use of the term *ambiguous* (SSA).

A similar situation arises in the case of quadrilaterals. The quadrilaterals are *not* congruent under these circumstances. If *A'E* is made equal to *A'D'* and *C'E* is made equal to *C'D'*, congruency *will* occur. Thus quadrilaterals need not be congruent even though four sides of one equal four sides of the other and a pair of corresponding angles are equal. Under the first set of circumstances the quadrilaterals will be non-convex (SSSSA).

Certain theorems which involve diagonals arise in the discussion of the congruency of quadrilaterals. These have no counterpart in the congruency of triangles, of course, since triangles have no diagonals. The problem of the 'floating diagonal' arises quite often in examining conditions under which quadrilaterals will be congruent.

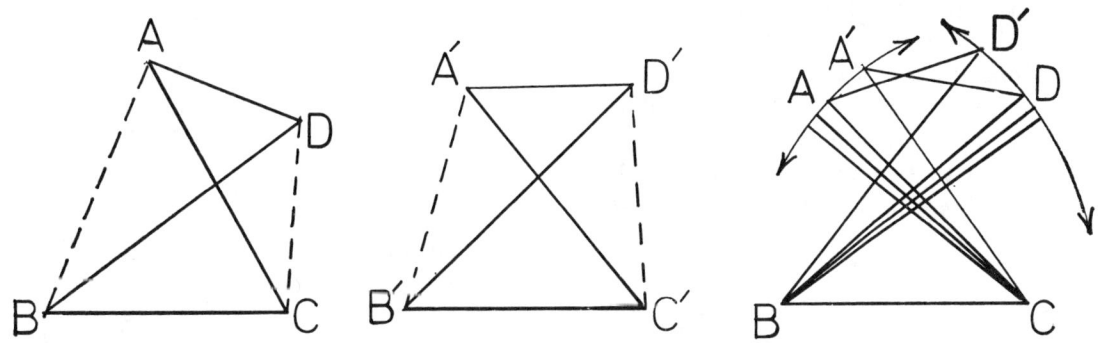

Quadrilaterals *ABCD* and *A'B'C'D'* are *not* necessarily congruent even though two sides and two diagonals are respectively equal. If *BC* is fixed and arcs the lengths of the diagonals are swung, side *AD* can be fitted in with varying orientations. Hence the term 'floating diagonal'. Such diagonals have to be fixed to assure rigidity and uniqueness.

Two quadrilaterals are congruent if a pair of corresponding sides and both pairs of diagonals are equal *and* the inclination of the diagonals to the corresponding sides is fixed.

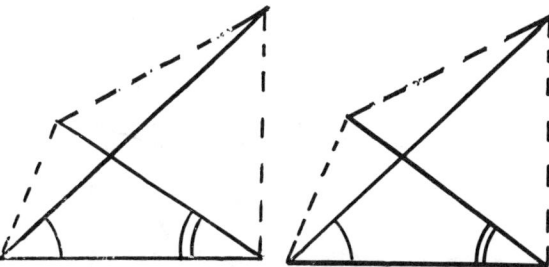

There are other general theorems which pertain to the congruence of quadrilaterals. Most productive are those pertinent to the congruence of special quadrilaterals: square, rectangle, parallelogram, rhombus, trapezium, isosceles trapezium, etc.

For polygons where the number of sides is greater than four, complications arise. A *complexity* of situations arises. These will test the mettle of your best prospective Einsteins.

Inventiveness in Geometry

An excellent example of enrichment for the many

A Tiny Treasury of TESSELLATIONS

BY ERNEST R. RANUCCI
State University of New York at Albany
Albany, New York

IF A series of points on some region of a plane is connected by segments in groups of three, the portion of the plane selected may be considered to be completely covered by the triangular regions formed. The same portion could, of course, be covered by a different set of triangular regions (Figs. 1 and 2).

Continuation of the process would ultimately result in the complete triangulation of the plane. The field of geometry concerned with two-dimensional space-filling is called tessellation. This is, technically, the process whereby a plane is completely covered by a set of nonoverlapping polygonal regions.

Among the more interesting of the tessellations are those which utilize one and only one type of regular polygon. There are but three of these tessellations. The square (vertex angle 90°), the equilateral triangle (vertex angle 60°), and the regular hexagon (vertex angle 120°) are the possibilities. The governing factor here requires the selection of just those regular polygons whose vertex angles divide 360° integrally. The regular pentagon (vertex angle 108°) could, obviously, not be used. (See Fig. 3.)

There are many ways of covering the plane with congruent nonregular polygonal regions and geometric configurations in general. Several of the more interesting possibilities are shown in the display on the opposite page.

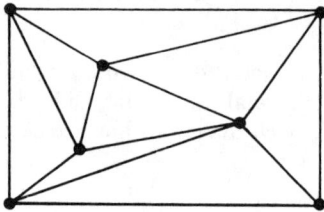

FIGURE 1

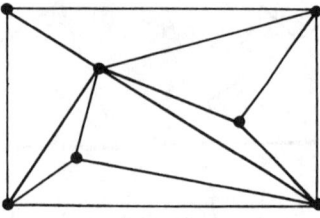

FIGURE 2

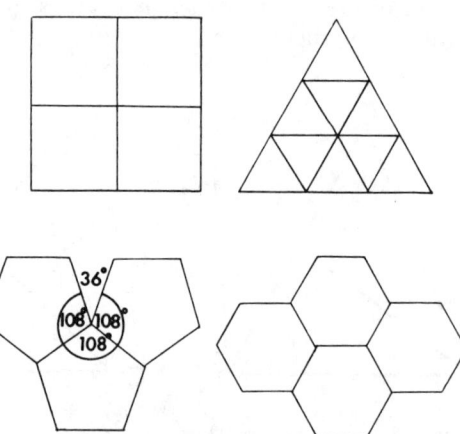

FIGURE 3

Imaginative Ideas: Ranucci's Reservoir

Inventiveness in Geometry

71

Any triangle, such as ABC in Figure 4, may be used to tessellate a plane. Construct lines through A and C parallel, respectively, to BC and AB.

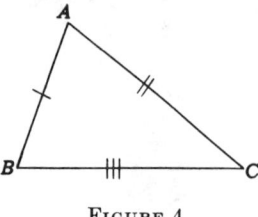

FIGURE 4

The resulting parallelogram, $ABCB'$ in Figure 5, then becomes the basic unit whose repetition will result in the complete tessellation of the plane. This so-called fundamental region consists, in reality, of the original triangular region and one congruent to it. Here, it is worth noting, A and C

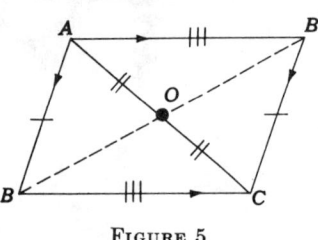

FIGURE 5

are symmetric with respect to O, the midpoint of AC. So are points B and B'. Both ABC and $AB'C$ are the same "face" of the original traingle. This is typical of symmetry with respect to a *point* and *not* typical of symmetry with respect to a line. Tessellation of a plane may, then, be accomplished by selecting any triangle, rotating it 180° in such a manner that two of the original edges coincide, and repeating the process. The fundamental parallelograms which result are, normally, of three types (Fig. 6).

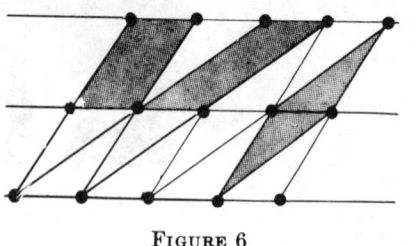

FIGURE 6

Each of the three edges of the original basic triangle becomes, in turn, a diagonal of one of the fundamental parallelograms. The vertices of the resulting tessellation are termed "lattice" points—a set of points having the property that a line joining any two of them contains not only these two, but infinitely many. Which of the points on the plane are closest to some particular lattice point—for example, A?

Construction of the perpendicular bisectors of AB, AD, AE, AF, AH, and AI may help to clarify the situation. Points closer to one lattice point than to another must lie on one or the other side of these perpendicular bisectors. The hexagonal cell which emerges answers the question. This hexagon—a Dirichlet region—is sometimes called a par-hexagon. Kasner and Newman define the par-hexagon as one whose opposite sides are not only equal but parallel. Opposite vertices of such a hexagon are always symmetric with respect to the center of the par-hexagon.

By repeating the Dirichlet hexagon of Figure 7 the entire plane may be covered.

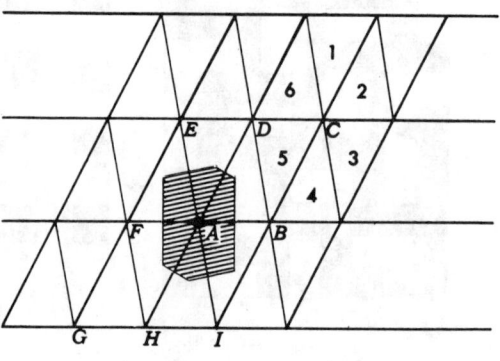

FIGURE 7

The polygonal region composed of Triangles 1, 2, 3, 4, 5, and 6 is also a Dirichlet region. It, too, may be used to "paper" the plane; it comes, however, from a different set of lattice points.

A further analysis of the par-hexagon shows that, surprisingly enough, *any* quadrilateral may be used in tessellation of

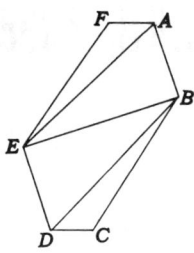

FIGURE 8

the plane. (See Fig. 8.) In par-hexagon $ABCDEF$

$$\triangle AFE \cong \triangle DCB$$
($AF = DC$ and $FE = CB$.
$\angle F = \angle C$, angles with
sides respectively parallel.)
$$\triangle ABE \cong \triangle DEB \text{ (s.s.s.)}$$
$\therefore$ Quad. $ABEF \cong$ Quad. $EDCB$.

This means that any quadrilateral, like those in Figure 9, may be used for a complete tessellation of the plane. The sum of

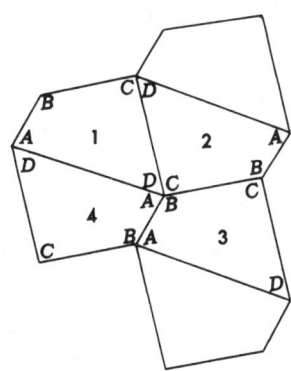

FIGURE 9

the angles about a point is 360°. So is the sum of the angles of any quadrilateral.

A quadrilateral need not even be convex

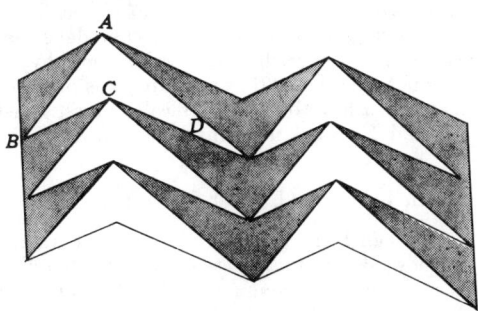

FIGURE 10

for two-dimensional space-filling. The reentrant or nonconvex polygon $ABCD$, if properly reversed and repeated, may be used for tessellation. Note that the vertices of the complex configuration of Figure 10 are still vertices of sets of parallelograms. Each double quadrilateral still encompasses a par-hexagon.

Figure 11 illustrates the case where pentagons may be used in tessellation of a

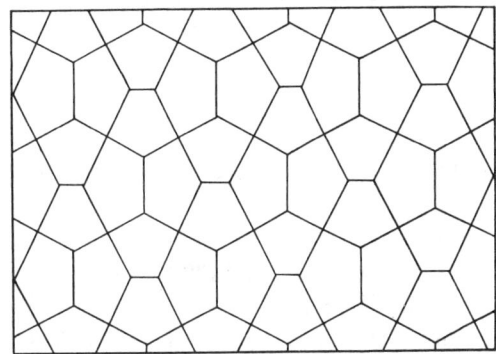

FIGURE 11

plane. The par-hexagons involved are a special type—actually, truncated rhombuses. The sides of the second set are perpendicular bisectors of the sides of the original set.

A unit on tessellations is excellent for enriching the customary material in a course in plane geometry on polygons. Although a great deal of material is available on tessellations with the regular polygons, the use of unorthodox configurations adds an extra flavor to such a unit.

BIBLIOGRAPHY

COXETER, H. S. M. *Introduction to Geometry.* New York: John Wiley & Sons, 1961.

HILBERT, D., and COHN-VOSSEN, S. *Geometry and the Imagination.* New York: Chelsea Publishing Co., 1956.

KASNER, EDWARD, and NEWMAN, JAMES R. *Mathematics and the Imagination.* New York: Simon & Schuster, 1940.

STEINHAUS, HUGO. *Mathematical Snapshots* (2nd ed.). New York: Oxford University Press, 1956.

WEBSTER, D. *Brain Boosters.* New York: The Natural History Press, 1966.

Inventiveness in Geometry

MASTER OF TESSELLATIONS: M. C. ESCHER, 1898-1972

An account of the contributions to geometry of a well-known artist.

By **ERNEST R. RANUCCI**

State University of New York at Albany
Albany, New York

ENGLAND lost its poet laureate in 1972—Cecil Day Lewis. If there is such a thing as an artist laureate, and there seems to be, those of us in the field of geometry lost ours in 1972 when Maurits Cornelis Escher died. Escher, a Dutch painter, was born in the Netherlands at Leeuwarden on 17 June 1898. His early training in art was traditional. Escher spent many years under the tutelage of S. Jessurun de Mesquita. From Mesquita he learned graphic techniques and became proficient in the art of the woodcut. Escher migrated to Italy in 1922, settled in Rome, and remained there until 1934. Plates 1 and 2 are typical of this period of Escher's work. The rendering of Italian landscapes and Italian buildings, particularly those of southern Italy, appealed to him. There is little here to suggest the Escher who was to come, the Escher of repetitive patterns and periodic drawings. Present, however, is the expertise of the master craftsman, the obvious command of his favorite medium—the woodcut.

Later Escher lived in Switzerland, then in Belgium, then in Holland. He finally settled in Baarn, the Netherlands, in 1941. Most of the rest of his life was spent there. The Escher most people know, the artist preoccupied with space filling of a repetitive nature, developed from an interest in the work of Moorish artists. The Moors occupied Spain from 711 to 1492. They were forbidden by their religion to depict animate objects. Escher says (1970, p. 11):

> This is the richest source of inspiration that I have ever struck; nor has it yet dried up ... a surface can be regularly divided into, or filled up with, similar-shaped figures (congruent) which are contiguous to one another, without leaving any open spaces. The Moors were past masters of this. They decorated walls and floors, particularly in the Alhambra in Spain, by placing congruent, multi-coloured pieces of majolica (tiles) together without leaving any spaces between. What a pity it is that Islam did not permit them to make "graven images." They always restricted themselves, in their massed tiles, to designs of an abstract geometrical type. Not one single Moorish artist, to the best of my knowledge, ever made so bold (or maybe the idea never occurred to him) as to use concrete, recognisable, naturalistically conceived figures of fish, birds, reptiles or human beings as elements in their surface coverage. This restriction is all the more unacceptable to me in that the *recognizability* of the components of my own designs is the reason for my unfailing interest in this sphere.

Plate 3 and plate 4, taken from notes of Escher, illustrate certain of the Moorish plane tessellations.

It is a curious fact that Escher's paintings should be so inextricably tied up with mathematics; he was no trained mathematician. In the preface to *Symmetry Aspects of M. C. Escher's Periodic Drawings*, used as a text in university courses in crystallography, Escher has this to say (Macgillavry, 1965):

> In the course of the years I designed about a hundred and fifty of these tessellations. In the beginning I puzzled quite instinctively, driven by an irresistible pleasure in repeating the same forms, without gaps, on a piece of paper. These first drawings were tremendously time-devouring because I had never heard of crystallography; so I did not even know that my game was based on rules which have been scientifically investigated. Nor had I visited the Alhambra at that time.

Again he says (Escher 1970, p. 9):

> Although I am absolutely innocent of training or knowledge in the exact sciences, I often seem to have more in common with mathematicians than with my fellow artists.

Escher, who made his own first periodic woodcut in 1922—a collection of eight different human heads—found in the Moorish approach the key to much of his life's work. (Plate 5) He simply adapted

Reproduction courtesy of the Vorpal Gallery, San Francisco

Plate 2. Castrovalva

Reproduction courtesy of the Vorpal Gallery, San Francisco

Plate 1. Coast of Amalfi

Inventiveness in Geometry

Plate 4. Study #2, tiles in the Alhambra

Plate 3. Study #1, tiles in the Alhambra

76 *Imaginative Ideas: Ranucci's Reservoir*

Moorish solutions, with a prodigious inventiveness of his own. He seems to have discovered on his own the seventeen fundamental ways of covering a plane through repetitive patterns. He applied axial symmetry, radial symmetry, elements of Poincaré's geometry; the list could go on and on. Another of his own comments on the fundamentally intuitive approach he had to his art is illuminating (Macgillavry 1965, p. viii):

> The dynamic action of making a symmetric tessellation is done more or less unconsciously. While drawing I sometimes feel as if I were a spiritualist medium, controlled by the creatures which I am conjuring up. It is as if they themselves decide on the shape in which they choose to appear. They take little account of my critical opinion during their birth and I cannot exert much influence on the measure of their development. They usually are very difficult and obstinate creatures.

Plates 6 and 7 illustrate other typical Escher tessellations.

Plates 8 and 9 illustrate still other Escher protean characteristics. I call this the *optical-illusion Escher*. His own comments on plate 8, *Concave and Convex*, bear repetition (Escher 1970, p. 11):

> Three little houses stand near one another, each under a cross-vaulted roof. We have an exterior view of the left-hand house, an interior view of the right-hand one and an either exterior or interior view of the one in the middle, according to choice. There are several similar inversions illustrated in this print; let us describe one of them. Two boys are to be seen, playing recorders. The one on the left is looking down through a

Reproduction courtesy of the Vorpal Gallery, San Francisco

Plate 5. Eight Heads

Inventiveness in Geometry

Reproduction courtesy of the Vorpal Gallery, San Francisco

Plate 6. Day and Night

Reproduction courtesy of the Vorpal Gallery, San Francisco

Plate 7. Fishes and Scales

window on to the roof of the middle house; if he were to jump down in front of it, he would land one story lower, on the dark coloured floor before the house. And yet the right-hand recorder player who regards that same cross-vault as a roof curving above his head will find, if he wants to climb out of *his* window that there is no floor for him to land on, only a fathomless abyss.

He has this to say about Plate 9, *Belvedere* (Escher 1970, p. 22):

In the lower left foreground there lies a piece of paper on which the edges of a cube are drawn. Two small circles marke the places where edges cross each other. Which edge comes at the front and which at the back? In a three-dimensional world simultaneous front and back is an impossibility and so cannot be illustrated. Yet it is quite possible to draw an object which displays a different reality when looked at from above and from below. The lad sitting on the bench has got just such a cubelike absurdity in his hands. He gazes thoughtfully at this incomprehensible object and seems oblivious to the fact that the belvedere* behind him has been built in the same impossible style. On the floor of the lower platform, that is to say indoors, stands a ladder which two people are busy climbing. But as soon as they arrive a floor higher they are back in the open air and have to re-enter the building. Is it any wonder that nobody in this company can be bothered about the fate of the prisoner in the dungeon who sticks his head through the bars and bemoans his fate?

* summer-house

The obituary of Escher that appeared in *Time* magazine of 10 April 1972 merely says this:

Died: Maurits Cornelis Escher, 73, Dutch artist known for his surrealistic woodcuts and lithography, in Hilversum, the Netherlands. Escher worked in almost complete obscurity for 30 years, until, in the early 30's, his vivid sense of fantasy and unusual uses of perspective won recognition in the U.S. His creations over half a century, about 270 works, now appear in museums on both sides of the Atlantic.

Like most obituaries, this one is too cold. Teachers of mathematics, especially those

Reproduction courtesy of the Vorpal Gallery, San Francisco

Plate 8. Concave and Convex

Inventiveness in Geometry

Reproduction courtesy of the Vorpal Gallery, San Francisco

Plate 9. Belvedere

with an interest in the spatial aspects of geometry, can well pay homage to this remarkable man by using what he gave the world. What better tribute to the artist than to include some of his work in the class in geometry?

For teachers with an interest in symmetry groups, books by Cadwell (1966, chap. 8) and Coxeter (1967, chap. 2) are recommended.

BIBLIOGRAPHY

Cadwell, J. H. *Topics in Recreational Mathematics.* Cambridge: At the University Press, 1966.

Coxeter, H. S. M. *Introduction to Geometry.* New York: John Wiley & Sons, 1967.

Escher, M. C. *The Graphic Work of M. C. Escher.* New York: Hawthorn Books, 1970.

———. *De Werelden van M. C. Escher.* Amsterdam: Meulenhoff International, 1971.

Macgillavry, Caroline H. *Symmetry Aspects of M. C. Escher's Periodic Drawings.* Utrecht: A. Oosthoek's Uitgeversmaatschappij NV, 1965.

Part 5

Games to Learn By

All of Ernest Ranucci's colleagues found him to be intrigued with games and puzzles of all types. He never worried about a rationale for the games. From his point of view, they were fun and math was fun, so that was rationale enough! Three readings have been included in this section. In the first, "What's in a Name?" from the *Grade Teacher,* he illustrated a game using students' names that would provide elementary students with calculation practice. In "Tantalizing Ternary," from the *Arithmetic Teacher,* he gave another approach to the well-known butcher problem and developed a game based on the ternary system. For classes studying bases other than ten, the problem and the game are a nonroutine application of one of these systems. In the final article in this section, "Dots and Squares," from the *Journal of Recreational Mathematics,* he analyzed the mathematical structure of a game and the winning strategies. The mathematics is not complex, and the many questions posed by the game could well be used as challenge problems for a bright senior high school class.

What's in a name?

Plenty! You can do away with arithmetic drill work and replace it with a fun game for your youngsters.

By ERNEST R. RANUCCI

ALL TOO OFTEN, students are given arithmetic drill work which is just plain tedium—an ordeal of adding column after column of figures with little, if any, immediate payoff.

Why subject your youngsters to that when you can make it fun by playing "Name Arithmetic"?

Few things are more familiar to a child than his name. So, to get your students' immediate interest, start with that.

On the chalkboard assign a value to each letter of the alphabet (see Figure 1). Ask each youngster to print his first and last names on a sheet of paper (no nicknames, please). Then ask each student to substitute the appropriate numbers for each letter in his name (see Figure 2). The next step is to add these number values to give each student the full numerical value of his name. For example, in Figure 2 the value of MARY CARPENTER is 157.

Use this introduction as the basis for the followup questions and activities:

1. Whose name gave the greatest value in the class? Whose name gave the least value in the class?

2. How many students have odd values for the sum of their coded letters? How many students have even values for the sum of their coded letters? Are these two results reasonably close? Is this to be expected?

3. How many students have odd values for both first and last names?

4. How many students have even values for both names?

5. How many students have an odd value for one of their names and an even value for the other?

6. How does the total number of students in Question 5 compare with that in Questions 3 and 4?

7. In how many cases does the value of the given name exceed that of the family name? In how many cases does the value of the family name exceed that of the given name? Which is more likely to occur? Does national origin have anything to do with this? Explain.

8. Reexamine the numerical values of the letters in your name. Write down an "O" for those which are odd and an "E" for those which are even (MARY CARPENTER; 13-1-18-25-3-1-18-16-5-14-20-5-18 becomes OOEO-OOEEOEEOE). Now proceed from left to right (that is, from letter to letter). Give yourself one point for every change in status—from odd to even (O to E) and from even to odd (E to O) count one point each; from even to even (E to E) and from odd to odd (O to O) count nothing. Accordingly, the value assigned to Mary Carpenter's name is 7. Who has the highest total in this game? Who has the lowest total? Is there anyone whose variations total zero? If not, could such a thing happen? (All numbers are even or all numbers are odd.)

9. Have each child select a word with an agreed-upon number of letters, perhaps five. The contest now is to see whose word has the highest coded total, the lowest coded total. Possible variations might include five-letter words with only odd numbers, only even numbers, etc.

This game can lead to all kinds of interesting possibilities and complications. Encourage creativ-

```
A B C D E F G H I
1 2 3 4 5 6 7 8 9

J K L M N O P Q R
10 11 12 13 14 15 16 17 18

S T U V W X Y Z
19 20 21 22 23 24 25 26
```
Figure 1

```
MARY    CARPENTER
13·1·18·25  3·1·18·16·5·14·20·5·18
        Total: 157
```
Figure 2

ity. For example, if a youngster comes up with *baa-ed* (making a five-letter word from the sound a sheep made), he's got a good low total of 13. If the student is that creative, accept it, by all means. The same goes for the youngster who comes up with *wuzzy* as in "Fuzzy-Wuzzy was a bear . . ." in a game to find the five-letter word with this highest value. Wuzzy totals 121, about as large a total as any youngster could hope for.

Note that many of these activities are only incidentally related to arithmetic, and in this lies their strength. Another advantage of these exercises is the fact that they involve fields other than mathematics—English and social studies, for example. The discovery and search for new words aids in vocabulary building. The investigation of family names may lead to fruitful learning. Activity 7 deals with the relative weights of family names and given names. Slavic names, for instance, are more likely to be heavy on the family name. Conversely, many Chinese family names are the two- or three-syllable types—Ho, Li, Eng, etc., and thus likely to be lighter on the family name.

Questions 3, 4, 5 and 6 have an aspect of probability connected with them. With regard to the nature (odd or even) of the values of the given name and the family name, there are four possibilities:

First Name	Last Name
Odd	Odd
Odd	Even
Even	Odd
Even	Even

This means that, ordinarily, about half the class will have names with like status (odd-odd or even-even) while half the class will have names with unlike status.

Encourage your students to come up with other arithmetic games based on their own names.

Mr. Ranucci is a professor of mathematics education at the State University of New York, Albany, N.Y.

Games to Learn By

Answers
"What comes next? Patterns in Figures" (p. 5)

14. E F G
15. V U T
16. 7 9 11
17. 23 27 31
18. E_9 F_{11} G_{13}
19. WD VE UF
20. R P N
21. 15 18 21
22. 16 22 29
23. 20 26 33
24. 2 32 2
25. 55555 666666 7777777
26. D23, E22, F21
27. 45 22½ 11¼
28. 12 14 16
29. P V C
30. 3 3½ 4
31. 16 19 22
32. 1/10 1/12 1/14
33. D Q E
34. 405 1215
35. 63 127
36. D T F
37. 4 6 5
38. I K M
39. H G J
40. 1 1/8 1 7/8 1 3/16
41. 3125 46,656
42. E
43. 15/16 21/22 28/29
44. seven
45. hat
46. DPW EQV
47. 1 lb. 11 oz.
48. LL
49. D_7 W_{20}
50. 31 30 31
51. 720 5040 40,320
52. !!!??? !!!!????
53. 28 33 38
54. 20 26 33
55. D W E
56. D T E
57. N K H
58. 2 ft. 3 in.
59. STUV
60. 34 55 89
61. 2 32 2
62. 10 9 12
63. O E P
64. 30 5

"Alphabet Soup" (p. 4)

(1) E   (7) X
(2) C   (8) Z
(3) I   (9) T
(4) R   (10) R
(5) V   (11) E
(6) ʃ   (12) O

*(All letters with curves in them are being listed alphabetically.)

Tantalizing ternary

ERNEST R. RANUCCI
State University of New York at Albany, Albany, New York

Dr. Ranucci is professor of mathematics education. Last summer he was in Colombia, South America, for three months on a Fulbright grant. His work there was with teacher education for both elementary and secondary levels.

Arithmetic computation in bases other than ten is a fertile field to explore. Especially interesting are certain problems that arise in the base three (ternary) system.

Place value in the ternary system is based upon powers of three. The units digit has a value of 1, the base digit a value of 3, the base-two digit a value of 9, the base-three digit a value of 27, etc. The number 12012 in base three has a value of 140 in base ten.

$$(12012_{three} = 140_{ten})$$

1	2	0	1	2
3^4	3^3	3^2	3^1	1
81	27	9	3	1

... Place value as a power of three
... Calculated place value

12012_{three}
$= 1 \times 81 + 2 \times 27 + 0 \times 9 + 1 \times 3 + 2 \times 1$
$= 81 \quad\quad 54 \quad\quad 0 \quad\quad 3 \quad\quad 2$
$= 140_{ten}.$

In the ternary system only three digits are needed. We shall use 0, 1, and 2.

Many intriguing puzzles are based on the ternary system. For the exploration of several of these, we need to convert the decimal numbers from 1 to 40 into base-three numerals. (See Table 1.) Note that

$$40 = 1 + 3 + 9 + 27.$$

This fact will prove to be of extreme significance later on.

Let us first consider the "butcher problem." A certain Scottish butcher finds that he can get along surprisingly well with

Table 1

Decimal notation	Ternary notation	Decimal notation	Ternary notation
1	1	21	210
2	2	22	211
3	10	23	212
4	11	24	220
5	12	25	221
6	20	26	222
7	21	27	1000
8	22	28	1001
9	100	29	1002
10	101	30	1010
11	102	31	1011
12	110	32	1012
13	111	33	1020
14	112	34	1021
15	120	35	1022
16	121	36	1100
17	122	37	1101
18	200	38	1102
19	201	39	1110
20	202	40	1111

nothing but a simple balance scale and individual weights of one, three, nine, and twenty-seven pounds. As long as no customer asks for meat in a quantity requiring the use of half-pounds, the butcher finds that he can weigh any integral number of pounds of meat from one to forty. He can, what's more, do this in one weighing (providing the customer is not so fastidious as to object to the placing of weights on the same pan with the cut of meat.) To weigh out five pounds of chopped beef, for example, he places a weight of nine pounds on one pan. He then places weights of one and three pounds on the other pan and adds meat until the scales balance (Fig. 1). This leads to the simple equation:

$$M + 1 + 3 = 9; M = 5.$$

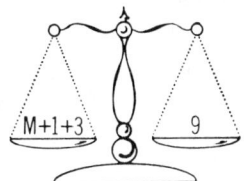

FIGURE 1

Weighing ten pounds of meat is easy. All this takes is the placing of weights of one and nine pounds on one pan and the addition of meat to the other pan until they balance (Fig. 2). The equation here is

$$M = 1 + 9; M = 10.$$

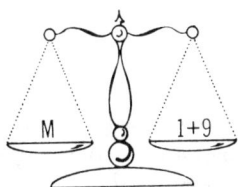

FIGURE 2

Curiously, each total, from one to forty, can be achieved by only one combination of the various weights available. Table 2, based on the ternary system, indicates the unique structure behind the "butcher problem." Remember that there is but one weight of each size.

The table is used as follows: All positive weights are placed on one pan; all negative weights are placed on the other pan. Meat is always added to the pan on which the negative weights have been placed. In weighing out 20 pounds, for example, weights of 27 and 3 are placed on one pan; weights of 9 and 1 are placed on the other. Meat is added to the 9, 1, side until a balance is secured. With the addition of another weight of 81 pounds to the set, every integral weight from 1 to 121 may be achieved. The "butcher" approach to equation solving can be handled by the average sixth grader if the approach is gradual.

Martin Gardner, in an article in the May 1964 issue of the *Scientific American*, offers

Table 2

1	1	21	27 − 9 + 3
2	3 − 1	22	27 − 9 + 3 + 1
3	3	23	27 − 3 − 1
4	3 + 1	24	27 − 3
5	9 − 3 − 1	25	27 − 3 + 1
6	9 − 3	26	27 − 1
7	9 − 3 + 1	27	27
8	9 − 1	28	27 + 1
9	9	29	27 + 3 − 1
10	9 + 1	30	27 + 3
11	9 + 3 − 1	31	27 + 3 + 1
12	9 + 3	32	27 + 9 − 3 − 1
13	9 + 3 + 1	33	27 + 9 − 3
14	27 − 9 − 3 − 1	34	27 + 9 − 3 + 1
15	27 − 9 − 3	35	27 + 9 − 1
16	27 − 9 − 3 + 1	36	27 + 9
17	27 − 9 − 1	37	27 + 9 + 1
18	27 − 9	38	27 + 9 + 3 − 1
19	27 − 9 + 1	39	27 + 9 + 3
20	27 − 9 + 3 − 1	40	27 + 9 + 3 + 1

an unusual approach to the "butcher problem." This starts with the value of the desired weight expressed in the ternary notation. To weigh out 33 pounds, for example, first consider the ternary numeral 1020_{three}. Add a 1 and subtract a 1 in the column where a 2 occurs. It is advantageous to rewrite as follows:

$$\begin{array}{r} \bar{1} \\ 1\,0\,2\,0 \\ 1 \end{array}$$

The addition of a 1 and the subtraction of a 1 does not affect the value of the number. (We are really adding 1×3 and subtracting 1×3, but there is no real need for this extra step.) The addition of 2 and 1 means that we can carry a 1 to the adjacent column. The result may now be expressed as $11\bar{1}0$. This means $27 + 9 - 3$, whose value is 33. Several solutions of this type are carried out:

20 pounds: 202_{three}
$$\begin{array}{r} \bar{1}\ \bar{1} \\ =\ 202 = 1\bar{1}1\bar{1} \\ 1\ 1 \end{array}$$

(This means $27 − 9 + 3 − 1 = 20$.)

32 pounds: 1012_{three}
$$\begin{array}{r} \bar{1} \qquad\quad \bar{1} \\ =\ 1012 = 102\bar{1} = 10\bar{2}\bar{1} = 11\bar{1}\bar{1} \\ 1 \qquad\quad 1 \end{array}$$

(This means $27 + 9 − 3 − 1 = 32$.)

23 pounds: 212_{three}
$$\begin{array}{r} \bar{1}\ \bar{1} \qquad\quad \bar{1} \\ =\ 212 = 1\bar{1}2\bar{1} = 1\bar{1}\bar{1}2\bar{1} = 10\bar{1}\bar{1} \\ 1\ 1 \qquad\quad 1 \end{array}$$

(This means $27 − 3 − 1 = 23$.)

Games to Learn By

The first forty numbers in the decimal system have values in the ternary system, numerals 1, 0, $\bar{1}$, as shown in Table 3.

Table 3

1	1	21	$1\bar{1}10$
2	$1\bar{1}$	22	$1\bar{1}11$
3	10	23	$10\bar{1}\bar{1}$
4	11	24	$10\bar{1}0$
5	$1\bar{1}\bar{1}$	25	$10\bar{1}1$
6	$1\bar{1}0$	26	$100\bar{1}$
7	$1\bar{1}1$	27	1000
8	$10\bar{1}$	28	1001
9	100	29	$101\bar{1}$
10	101	30	1010
11	$11\bar{1}$	31	1011
12	110	32	$11\bar{1}\bar{1}$
13	111	33	$11\bar{1}0$
14	$1\bar{1}\bar{1}\bar{1}$	34	$11\bar{1}1$
15	$1\bar{1}\bar{1}0$	35	$110\bar{1}$
16	$1\bar{1}\bar{1}1$	36	1100
17	$1\bar{1}0\bar{1}$	37	1101
18	$1\bar{1}00$	38	$111\bar{1}$
19	$1\bar{1}01$	39	1110
20	$1\bar{1}1\bar{1}$	40	1111

Most puzzles of the type, "Find your age on each of the following cards and I'll tell you what it is," are based on the ternary system. So are a series of puzzles in which names are guessed. Following is a description of one of the better of the name puzzles.

First prepare four cards according to the directions which follow.

GIRL'S NAME PUZZLE

First card (Worth 27 points)

(Front) Brunette, Brown Eyes

Dinah	Florence	Helen
Dolores	Fredrika	Hilda
Dorothy	Gail	Ida
Edith	Geraldine	Ilene
Elizabeth	Gertrude	Ingeborg
Eloise	Grace	Ingrid
Ethel	Gwendolyn	Jacqueline
Fanny	Harriet	Jeannette
Flora	Hazel	Joy

(Back) Brunette, Blue Eyes

Nancy	Petunia	Ursula
Naomi	Rita	Vicki
Nina	Roberta	Viola
Nora	Sarah	Wanda
Ocelie	Suzanne	Wendy
Ophelia	Tallulah	Winifred
Pamela	Tina	Yvette
Patsy	Tess	Yvonne
Paula	Tracy	Zaza

Second card (Worth 9 points)

(Front) Blonde, Brown Eyes

Barbara	Joy	Nancy
Betsy	Helen	Naomi
Betty	Hilda	Nina
Bunny	Ida	Nora
Carmen	Ilene	Ocelie
Carol	Ingeborg	Ophelia
Charlotte	Ingrid	Pamela
Corinne	Jacqueline	Patsy
Diane	Jeannette	Paula

(Back) Blonde, Blue Eyes

Dinah	Lana	Ursula
Dolores	Lena	Vicki
Dorothy	Lisette	Viola
Edith	Louise	Wanda
Elizabeth	Mary	Wendy
Eloise	Margaret	Winifred
Ethel	Mildred	Yvette
Fanny	Marilyn	Yvonne
Flora	Nanette	Zaza

Third card (Worth 3 points)

(Front) Auburn, Brown Eyes

Alice	Gwendolyn	Nancy
Anne	Harriet	Naomi
Audrey	Hazel	Nina
Charlotte	Jacqueline	Petunia
Corinne	Jeannette	Rita
Diane	Joy	Roberta
Ethel	Lana	Ursula
Fanny	Lena	Vicki
Flora	Lisette	Viola

(Back) Auburn, Blue Eyes

Barbara	Helen	Pamela
Betsy	Hilda	Patsy
Betty	Ida	Paula
Dinah	Katherine	Tina
Dolores	Kay	Tess
Dorothy	Kyle	Tracy
Florence	Marilyn	Yvette
Fredrika	Mildred	Yvonne
Gail	Nanette	Zaza

Fourth card (Worth 1 point)

(Front) Red, Brown Eyes

Abigail	Grace	Nancy
Audrey	Hazel	Nora
Betty	Ida	Pamela
Carol	Ingrid	Petunia
Diane	Joy	Sarah
Dorothy	Katherine	Tina
Eloise	Lana	Ursula
Flora	Louise	Wanda
Gail	Mildred	Yvette

(Back) Red, Blue Eyes

Alice	Gwendolyn	Nina
Barbara	Helen	Ophelia
Bunny	Ilene	Paula
Charlotte	Jacqueline	Roberta
Dinah	Joyce	Tallulah
Edith	Kyle	Tracy
Ethel	Lisette	Viola
Florence	Margaret	Winifred
Geraldine	Nanette	Zaza

Following is a list of possible names, with numerical value.

Positive

1. Abigail		21. Fanny
2. Alice		22. Flora
3. Anne		23. Florence
4. Audrey		24. Fredrika
5. Barbara		25. Gail
6. Betsy		26. Geraldine
7. Betty		27. Gertrude
8. Bunny		28. Grace
9. Carmen		29. Gwendolyn
10. Carol		30. Harriet
11. Charlotte		31. Hazel
12. Corinne		32. Helen
13. Diane		33. Hilda
14. Dinah		34. Ida
15. Dolores		35. Ilene
16. Dorothy		36. Ingeborg
17. Edith		37. Ingrid
18. Elizabeth		38. Jacqueline
19. Eloise		39. Jeannette
20. Ethel		40. Joy

Negative

$^-$1. Joyce	$^-$21. Patsy
$^-$2. Katherine	$^-$22. Paula
$^-$3. Kay	$^-$23. Petunia
$^-$4. Kyle	$^-$24. Rita
$^-$5. Lana	$^-$25. Roberta
$^-$6. Lena	$^-$26. Sarah
$^-$7. Lisette	$^-$27. Suzanne
$^-$8. Louise	$^-$28. Tallulah
$^-$9. Mary	$^-$29. Tina
$^-$10. Margaret	$^-$30. Tess
$^-$11. Mildred	$^-$31. Tracy
$^-$12. Marilyn	$^-$32. Ursula
$^-$13. Nanette	$^-$33. Vicki
$^-$14. Nancy	$^-$34. Viola
$^-$15. Naomi	$^-$35. Wanda
$^-$16. Nina	$^-$36. Wendy
$^-$17. Nora	$^-$37. Winifred
$^-$18. Ocelie	$^-$38. Yvette
$^-$19. Ophelia	$^-$39. Yvonne
$^-$20. Pamela	$^-$40. Zaza

The puzzle is worked as follows: Someone is asked to pick a name from the cards and to identify the hair color and eyes each time the name is mentioned. The score on any card is considered positive when the eyes are brown. The score is considered negative when the eyes are blue. Thus: Rita—Brunette, Blue Eyes ($^-$27); Auburn, Brown Eyes (3). Add the scores algebraically. The sum is $^-$24. Look up $^-$24 in the list of names and identify "Rita."

Thus: Joy (27, 9, 3, 1) or 40
Yvonne ($^-$27, $^-$9, $^-$3) or $^-$39
Harriet (27, 3) or 30
Ethel (27, $^-$9, 3, $^-$1) or 20

The structure of the Girl's Name Puzzle is identical to that of the "butcher problem." Once the desired number has been converted to the Gardner positive-negative number, the name associated with that number is placed on the appropriate face of the proper card.

An examination of other number systems will reveal the unique nature of the ternary system. Suppose that the weights used in the "butcher problem" had been 1, 2^1, 2^2, 2^3, etc.—elements of the binary system. Then meat, in the amounts indicated below, could have been weighed as follows:

1. 1 or 4 − 2 − 1, etc.
2. 2 or 8 − 4 − 2, etc.
3. 1 + 2 or 4 − 1, etc.
4. 4 or 8 − 4, etc.

Such weights could, of course, be used, but there would be nothing unique about such a system. Each weight could be produced in a variety of ways.

A base-four (quaternary) system presents its own problems. Let us attempt to weigh out the first twenty pounds; we may use only individual weights of 1, 4, 16, 64, etc.

1. 1	11. 16 − 4 − 1
2. ?	12. 16 − 4
3. 4 − 1	13. 16 − 4 + 1
4. 4	14. ?
5. 4 + 1	15. 16 − 1
6. ?	16. 16
7. ?	17. 16 + 1
8. ?	18. ?
9. ?	19. 16 + 4 − 1
10. ?	20. 16 + 4

Certain "clusters" appear with a predictable pattern, but many "holes in the Swiss cheese" show up. This deficiency becomes aggravated when the number base increases.

It would appear that the ternary system alone allows us to achieve all integral weights in a unique manner.

Games to Learn By

Dots and Squares

Ernest R. Ranucci
State University of New York
Albany, New York

One of the identifying characteristics of the contemporary mathematician seems to be his insatiable curiosity about anything or everything. Anything or everything, that is, which seems to have underlying mathematical principles behind its operation. A group of graduate students at Newark State College, Union, New Jersey, at the suggestion of Gail Koplin, got interested in the game called *Dots and Squares* (or, sometimes, just *Dots*). A commercial version of the game is called *Square It*. This is the game consisting of a square array of points and is customarily played by two people. Each, in turn, connects a pair of points until such time as a square is enclosed. The person enclosing a square places his initials in the square and the game continues until all the squares are claimed by one or the other contestant. Various strategies occur during the game. Sometimes the closing of the fourth side of a square then presents opportunities for the closing of a whole string of squares. Sometimes the sacrifice of one group of squares to the opponent places *him* in the unenviable position of having to sacrifice a greater number of squares than he received.

There are two versions of the game. One version calls for strict alternate play throughout; the other gives a player an extra turn or stroke after the completion of a square. A winning strategy for the second player if the first version is played on a board with an even number of dots on a side is to play symmetrically opposite the stroke made by the first player. The second player will win with one square more than the first player. An analysis of the strategy for the second version of the game is given in [1] and [2].

In this paper we are not concerned with winning strategies, but only with the game up to the point just before squares are completed. Both versions of the game are equivalent to this point. The group at Newark State College limited itself to square arrays of points and sought to solve just two problems:

1. What is the greatest number of moves that may be made before one of the contestants is presented with the opportunity for completion of at least one square. (This is analogous to the maneuver in checkers where a contestant *must* jump an opponent if given the opportunity.)

2. What is the greatest number of moves that may be made before the next move of one of the contestants *forces* the completion of at least one square.

Both of the problems seem to be related to an odd-even relationship. Let n represent the number of points on one side of the square array. Consider first the case where the number of points on a side is odd (Figure 1).

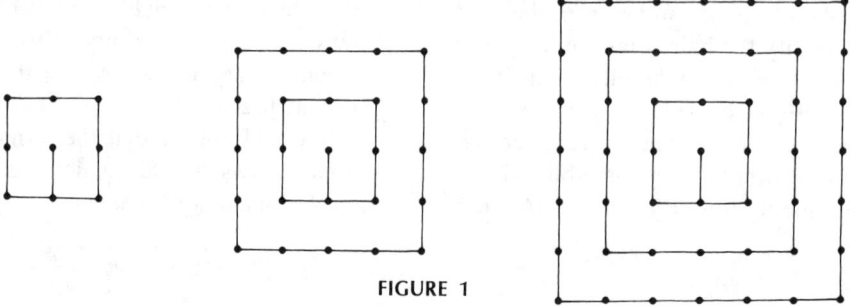

FIGURE 1

The number of moves can be predicted through the value n^2 when n is odd. The preliminary maneuvers in "ringing" the squares (we called it "Ring Around the Rosy") seem to postpone the inevitable moment when a movement must complete a square. Consider the even case (Figure 2). The number of moves may be predicted through the value $n^2 - 1$.

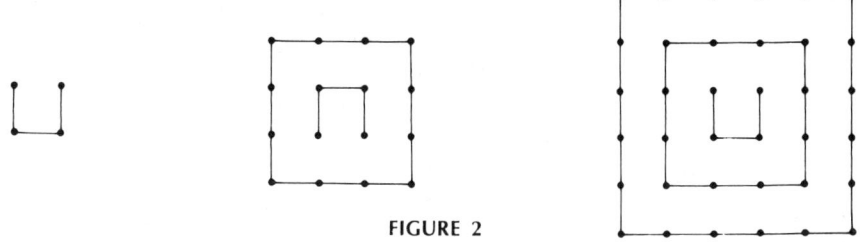

FIGURE 2

With respect to problem 2, two cases again arise. When the number of points on a side is odd, the number of moves required is as indicated in Figure 3.

Consider the series involving the numbers encountered: 10, 32, 66, 112. The first and second differences are as follows:

$$\begin{array}{cccc} 10 & 32 & 66 & 112 \\ & 22 & 34 & 46 \\ & & 12 & 12 \end{array}$$

The constant second difference suggests that the function involved is quadratic. The general quadratic $ax^2 + bx + c$ has values as follows when x is replaced in turn by 1, 2, 3, and 4:

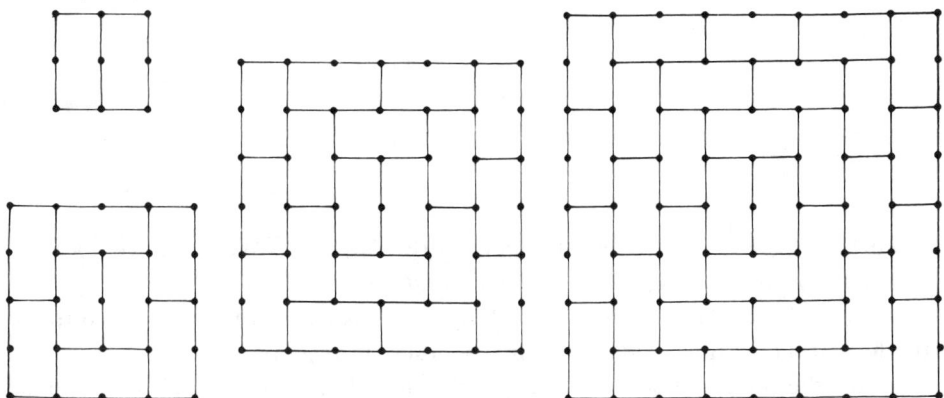

FIGURE 3

$$\begin{array}{cccc} a+b+c & 4a+2b+c & 9a+3b+c & 16a+4b+c \\ & 3a+b & 5a+b & 7a+b \\ & & 2a & 2a \end{array}$$

In the quadratic function we seek $a = 6$, $b = 4$, and $c = 0$. (For a complete explanation of this valuable quadratic technique see [3].) The quadratic function in our case is, then, $6x^2 + 4x$. Just how is x related to n (the number of points on the side of the square)? Let $2x + 1 = n$. Then the following table results:

n	x	$6x^2 + 4x$
3	1	10
5	2	32
7	3	66
9	4	112
11	5	170

Games to Learn By

When the number of points on a side is even, the number of moves required would be as shown in Figure 4.

By the application of the quadratic technique outlined before, the function is seen to be $6x^2 - 2x - 1$.

$$\begin{array}{cccc} 3 & 19 & 47 & 87 \\ & 16 \quad 28 \quad 40 & & \\ & 12 \quad 12 & & \end{array}$$

With $a = 6, b = -2, c = -1$.

Here x is related to n through the equation $2x = n$:

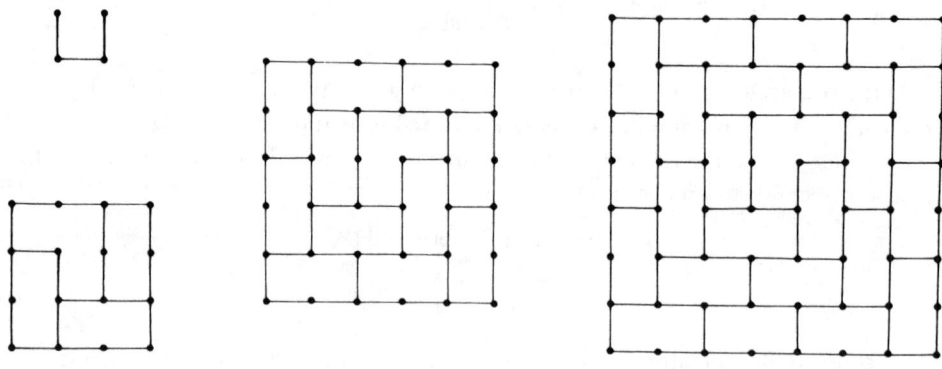

FIGURE 4

n	x	$6x^2 - 2x - 1$
2	1	3
4	2	19
6	3	47
8	4	87

There are many interesting off-shoots to this problem. We invite cerebration on the following aspects of the *Dots and Squares* game:

1. Can it be shown that the "Ring Around the Rosy" maneuver will postpone the formation of squares longer than any other strategem?

2. Are the results of this study applicable to other arrays of points, irregular in nature?

3. The vertices of a cube represent a cubical array of points. A cube has twelve edges; if only eleven of these are joined, the cube will be incomplete. What are the principles behind the completion of cubical arrays of points?

4. How are the rules of this game affected if the shape of the array of admissible cells is equilateral? Pentagonal? Hexagonal? A combination of regular octagons and alternately placed squares?

References

1. Thomas S. Briggs, "An Analysis of 'Square It,'" *Recreational Mathematics Magazine*, No. 4 (August 1961), pp. 52-54.
2. J. C. Hollady, "A Note on the Game of Dots," *American Mathematical Monthly*, Vol. 73, No. 7 (Aug-Sep 1966), pp. 717-720.
3. W. W. Sawyer, "A Method of Discovery," *The Mathematics Student Journal*, Vol. 6, No. 1 (November 1958), pp. 4-5.

Selected Bibliography of Works by Ernest R. Ranucci

The following chronological list does not contain publications in Spanish and is missing some of the less accessible sources that are not referenced in standard guides.

Articles

"Applications." *Mathematics Teacher* 46 (December 1953): 567.

"His Designs Come from Math Books." *Popular Science* 168 (May 1956): 205–8.

"Weequahic Configuration." *Mathematics Teacher* 53 (February 1960): 124–26.

"What Comes Next? Patterns in Figures." *Instructor* 73 (September 1963): 40.

"4 Areas in the New Math." *Instructor* 73 (February 1964): 62–65.

"Discovery in Mathematics." *Arithmetic Teacher* 12 (January 1965): 14–18.

"Function Follows Form." *Arithmetic Teacher* 13 (April 1966): 278–82.

"P . . . A . . . V. . . ." *New York State Mathematics Teachers Journal* 16 (1966): 33–37.

"Anyone for Tennis?" *School Science and Mathematics* 67 (December 1967): 761–65.

"Jungle-Gym Geometry." *Mathematics Teacher* 61 (January 1968): 25–28.

"A Tiny Treasury of Tessellations." *Mathematics Teacher* 61 (February 1968): 114–17.

"Alphabet Soup." *Jack and Jill,* September 1968, pp. 44, 68.

"The Four Color Game." *Grade Teacher* 86 (October 1968): 109–10.

"Tantalizing Ternary." *Arithmetic Teacher* 15 (December 1968): 718–21.

"Geometry and the Animal World." *Primary Mathematics* 6, no. 1 (1968): 48–49.

"Spatial Aspects of the Venn Diagram." *New York State Mathematics Teachers Journal* 18 (1968): 64–67.

"Dots and Squares." *Journal of Recreational Mathematics* 2 (January 1969): 57–60.

"Tips for the Beginning Teacher." *New Jersey Mathematics Teacher* 4 (May 1969): 5–11.

"What's in a Name?" *Grade Teacher* 87 (January 1970): 48, 50.

"On Skewed Regular Polygons." *Mathematics Teacher* 63 (March 1970): 219–22.

"The Alpha-Omega of Geometric Phenomena." *New York State Mathematics Teachers Journal* 20 (1970): 127–29.

"Aspects of Combinatorial Geometry." *School Science and Mathematics* 70 (April 1971): 338–44.

"Schlegel Diagrams." *Journal of Recreational Mathematics* 4 (April 1971): 106–13.

"Curves from Polygons." *Mathematics Teaching* 55 (Summer 1971): 10–12.

"Space-Filling in Two Dimensions." *Mathematics Teacher* 64 (November 1971): 587–93.

"Teaching Permutations." *Grade Teacher* 89 (November 1971): 54–55, 77.

"Gnomons." *Connecticut Mathematics Journal* 4 (December 1971): 5–8.

"Permutation Patterns." *Mathematics Teacher* 65 (April 1972): 333–38.

"Topology through the Alphabet." *Mathematics Teacher* 65 (December 1972): 687–90.

"Constructive Constructions." *New York State Mathematics Teachers Journal* 22, no. 4 (1972): 164–67.

"Music in the Marshall Islands." *School Science and Mathematics* 73 (April 1973): 319–26.

"The Congruency of Quadrilaterals." *Mathematics Teaching* 64 (September 1973): 35–37.

[Coauthored with Margaret Farrell] "On the Occasional Incompatibility of Algebra and Geometry." *Mathematics Teacher* 66 (October 1973): 491–97.

"Cutting Candles." *Mathematics in School* 2 (November 1973): 24–25.

"The Number Game." *New York State Mathematics Teachers Journal* 23, no. 2 (1973): 88.

"Board to Death." *New York State Mathematics Teachers Journal* 23, no. 3 (1973): 133.

"Convex Complexities." *New York State Mathematics Teachers Journal* 23, no. 4 (1973): 162–63.

"Fruitful Mathematics." *Mathematics Teacher* 67 (January 1974): 5–14.

"Master of Tessellations: M. C. Escher, 1898–1972." *Mathematics Teacher* 67 (April 1974): 299–306.

"Round and Round." *Mathematics Student* 21 (April 1974): 1–2.

"Learn a New Game." *Instructor* 83 (June 1974): 48.

"The Alpha-Omega of Geometric Phenomena, Part 2." *New York State Mathematics Teachers Journal* 24, no. 3 (1974): 116–19.

"Of Shoes—and Ships—and Sealing Wax—of Barber Poles and Things." *Mathematics Teacher* 68 (April 1975): 261–64.

"On the Steepness of Cones." *Mathematics Teacher* 69 (February 1976): 140

"Isosceles^n." *Mathematics Teacher* 69 (April 1976): 289–94.

"Circular Tangrams from 6 to 16." *Instructor* 85 (May 1976): 68, 70.

"Design à la Escher." *Design* 77 (Midsummer 1976): 14–15.

"Penny Raffle." *New York State Mathematics Teachers Journal* 26, no. 2 (1976): 21–28.

"Blockbusters." *New Jersey Mathematics Teacher* 34, no. 2 (1977): 5–12.

"Snipshots." *Arithmetic Teacher* 25 (February 1978): 40–41.

"The World of Buckminster Fuller." *Mathematics Teacher* 71 (October 1978): 568–77.

Books

Points, Lines, and Planes: An Introduction to Geometry in Two Dimensions. New York: Macmillan, 1963.

Four by Four. Boston: Houghton Mifflin Co., 1968.

Tessellation and Dissection. Portland, Me.: J. Weston Walch, Publisher, 1970.

Georule Activities. Palo Alto, Calif.: Creative Publications, 1971

[Coauthored with J. L. Teeters] *Creating Escher-Type Drawings.* Palo Alto, Calif.: Creative Publications, 1977.

[Coauthored with W. E. Rollins] *Curiosities of the Cube.* New York: Thomas Y. Crowell Co., 1977.

[Coauthored with W. E. Rollins] *Brain Drain.* Fresno, Calif.: Creative Teaching Associates, 1978.

Book Chapters

"The Calculus of Finite Differences: Enrichment for Student and Teacher." In *Professional Growth for Teachers: Mathematics: Junior High School Edition,* edited by Gerald R. Rising. New London, Conn.: Croft Educational Services, April 1965.

"A Colorful Approach to Mathematics: Content for Discovery Lessons." In *Professional Growth for Teachers: Mathematics: Junior High School Edition,* edited by Gerald R. Rising. New London, Conn.: Croft Educational Services, November 1964.

"Developing Spatial Imagination." In *Updating Mathematics: Junior High School Teachers Edition,* edited by Francis J. Mueller, pp. 159–62. New London, Conn.: Croft Educational Services, 1964.

"Drawing for Math Teachers Who Can't Draw." In *Updating Mathematics: High School Teachers Edition,* edited by Francis J. Mueller, pp. 49–52. New London, Conn.: Croft Educational Services, 1964.

"Introduction to Symmetry." In *Updating Mathematics: Junior High School Teachers Edition,* edited by Francis J. Mueller, pp. 155–58. New London, Conn.: Croft Educational Services, 1964.

"Topology of the Alphabet." In *Updating Mathematics: High School Teachers Edition,* edited by Francis J. Mueller, pp. 115–24. New London, Conn.: Croft Educational Services, 1964.